Bounded Cohomology and Simplicial Volume

Since its introduction by Gromov in the 1980s, the study of bounded cohomology and simplicial volume has developed into an active field connected to geometry and group theory. This monograph, arising from a learning seminar for young researchers working in the area, provides a collection of different perspectives on the subject, both classical and recent. The book's introduction presents the main definitions of the theories of bounded cohomology and simplicial volume, outlines their history, and explains their principal motivations and applications. Individual chapters then present different aspects of the theory, with a focus on examples. Detailed references to foundational papers and the latest research are given for readers wishing to dig deeper. The prerequisites are only basic knowledge of classical algebraic topology and of group theory, and the presentations are gentle and informal in order to be accessible to beginning graduate students wanting to enter this lively and topical field.

CATERINA CAMPAGNOLO is a postdoctoral researcher now working at UAM Madrid.

FRANCESCO FOURNIER-FACIO is a PhD student at ETH Zürich.

NICOLAUS HEUER received his PhD from the University of Oxford.

MARCO MORASCHINI is a type A fixed-termed Researcher at the University of Bologna. He was previously a postdoctoral researcher at the University of Regensburg.

LONDON MATHEMATICAL SOCIETY LECTURE NOTE SERIES

Managing Editor: Professor Endre Süli, Mathematical Institute, University of Oxford, Woodstock Road, Oxford OX2 6GG, United Kingdom

The titles below are available from booksellers, or from Cambridge University Press at www.cambridge.org/mathematics

Bounded Cohomology and Simplicial Volume

Edited by

CATERINA CAMPAGNOLO
Autonomous University of Madrid

FRANCESCO FOURNIER-FACIO
ETH Zürich

NICOLAUS HEUER

MARCO MORASCHINI
University of Bologna

CAMBRIDGE
UNIVERSITY PRESS

Shaftesbury Road, Cambridge CB2 8EA, United Kingdom

One Liberty Plaza, 20th Floor, New York, NY 10006, USA

477 Williamstown Road, Port Melbourne, VIC 3207, Australia

314–321, 3rd Floor, Plot 3, Splendor Forum, Jasola District Centre,
New Delhi – 110025, India

103 Penang Road, #05–06/07, Visioncrest Commercial, Singapore 238467

Cambridge University Press is part of Cambridge University Press & Assessment,
a department of the University of Cambridge.

We share the University's mission to contribute to society through the pursuit of
education, learning and research at the highest international levels of excellence.

www.cambridge.org
Information on this title: www.cambridge.org/9781009183291
DOI: 10.1017/9781009183284

First published 2023

A catalogue record for this publication is available from the British Library

ISBN 978-1-009-18329-1 Paperback

Contents

Contributors

Filippo Baroni *Mathematical Institute, University of Oxford, United Kingdom*

Caterina Campagnolo *Institute of Mathematical Sciences, Madrid, Spain*

Lizhi Chen *School of Mathematics and Statistics, Lanzhou University, P.R. China*

Francesco Fournier-Facio *Department of Mathematics, ETH Zürich, Switzerland*

Anton Hase *Department of Mathematics, Ben-Gurion University of the Negev, Israel*

Nicolaus Heuer *Unaffiliated*

Holger Kammeyer *Mathematical Institute, Heinrich Heine University Düsseldorf, Germany*

Biao Ma *Department of Mathematics, University of Côte d'Azur, France*

Marco Moraschini *Faculty of Mathematics, University of Regensburg, Germany*

Filippo Sarti *Department of Mathematics, University of Bologna, Italy*

Alessio Savini *Section of Mathematics, University of Geneva, Switzerland*

Shi Wang *Department of Mathematics, Michigan State University, United States of America*

Preface

The present volume consists of the proceedings of the *International Young Seminar on Bounded Cohomology and Simplicial Volume* held online from November 2020 to February 2021. This series of twelve talks was intended to be a gentle introduction to the main problems around the theory of bounded cohomology and simplicial volume as well as to many of their applications. Especially in this period in which travelling was very difficult (if not impossible), we founded this event as an opportunity to keep in touch with all the young researchers in the area and to provide both Master's and PhD students with an alternative to usual winter schools. Inspired by the style of the *"What is ... "* articles in the *Notices of the American Mathematical Society*, we invited young researchers to answer this question about the specific area of the subject they are working on. Via their 45-minute talks, we gathered different perspectives on bounded cohomology and simplicial volume, as well as related invariants.

As a complementary source for the audience, we asked all the speakers to kindly write a short report on their own talk: these reports are collected here. They are meant to be an introductory reading for those who are interested in learning the main ideas and definitions in the theory. Indeed, the emphasis is put on giving precise definitions, always accompanied by many examples, and theorems, together with the main ideas behind the proofs.

With this collection we aim to offer the reader a convenient starting point to learn about the foundational results of simplicial volume and bounded cohomology, as well as about the newer developments, in a gentle and informal way. Plenty of references should then point toward the relevant literature for readers interested in a further study of particular topics. A recurring reference in this collection is the book *Bounded Cohomology of Discrete Groups* by Roberto Frigerio [Frigerio, 2017]. The reader may be wondering how this collection diverges from it. We think that the greatest difference between the two books lies in their scope. Indeed, our aim was not to repeat the theory of simplicial volume and bounded cohomology, already treated in a very complete and satisfactory manner by Frigerio. Rather, the purpose of this collection is to offer accessible introductions to those many subjects gravitating around and emerging from the classical theory. In fact, the exhaustive approach by Frigerio had the consequence that most of the recent developments in the theory collected here are only briefly mentioned as *further readings* there.

How to Read This Book

Every chapter, as coming from an individual talk, is mainly self-contained. However, given the recurring audience, the new talks were allowed to use the material from the previous ones. Consequently, this might be true also for some chapters of this collection. When this is the case, internal references are provided.

We advise a reader beginning in the subject to read the introduction before starting with the individual chapters: this should provide the necessary background and notation for the sequel. The first half of the book primarily deals with *spaces*, geometric aspects and simplicial volume, while the second half is more concerned with *groups*, algebraic aspects and bounded cohomology. Of course both points of view are intertwined.

Overview

Among the results in this collection, we present two of the main theorems in the theory. Namely, we give a proof of Gromov's Mapping Theorem in Chapter 1. Then, Chapters 2 and 12 give two different proofs of the Gromov–Thurston Proportionality Principle. The familiarity with bounded cohomology developed through this book should be helpful to appreciate the second, more algebraic proof. The main differential geometric aspects in the theory are further discussed in Chapters 3 and 4, via the interactions of simplicial volume, curvature, and systoles. A more topological approach is taken in Chapter 5 by explaining the interplay of ergodic theory and simplicial volume. This direction is further developed in Chapter 6 about ℓ^2-invariants. The transition from geometry to algebra happens in Chapter 7, where stable commutator length is introduced together with applications to simplicial volume. Bavard Duality (Definition 7.6) connects stable commutator length to quasimorphisms: Chapters 8 and 9 describe how to construct these for classes of negatively curved groups. Finally, we leave the discrete world to explore the bounded cohomology of topological groups. To this end, symmetric spaces are introduced in Chapter 10 along with the Dupont conjecture. Moreover, some applications of continuous bounded cohomology to rigidity theory are discussed in Chapter 11.

Acknowledgments

The editors warmly thank the University of Regensburg for virtually hosting the seminars, as well as the grant CRC 1085 *Higher Invariants* (University of Regensburg funded by DFG) for its support.

The editors are also in debt to Roberto Frigerio, Clara Löh, Alessandra Iozzi, Maria Beatrice Pozzetti, Roman Sauer, and Alessandro Sisto for many useful comments on the first draft of this collection.

Introduction

Simplicial Volume and Basic Properties

Simplicial volume is a homotopy invariant of manifolds introduced in Gromov's proof of Mostow rigidity [Munkholm, 1980]. It measures the complexity of a manifold in terms of the singular chains representing its real fundamental class. More precisely, given a topological space X we define the ℓ^1-*norm* on the space of real singular n-chains $C_n(X; \mathbb{R})$ as follows:

$$\left\| \sum_{i=1}^{k} a_i \sigma_i \right\|_1 = \sum_{i=1}^{k} |a_i|$$

for all $\sum_{i=1}^{k} a_i \sigma_i \in C_n(X; \mathbb{R})$ in reduced form. The ℓ^1-norm then descends to a seminorm, called ℓ^1-*seminorm*, when we consider homology groups:

$$\|\alpha\|_1 = \inf \left\{ \left\| \sum_{i=1}^{k} a_i \sigma_i \right\|_1 \ \middle| \ \left[\sum_{i=1}^{k} a_i \sigma_i \right] = \alpha \right\}$$

for all $\alpha \in H_n(X; \mathbb{R})$. This leads to the definition of simplicial volume:

Definition 1 Let M be a closed connected oriented topological manifold of dimension n. The *simplicial volume* of M is defined to be the ℓ^1-seminorm of its real fundamental class $[M] \in H_n(M; \mathbb{R})$, that is,

$$\|M\| := \|[M]\|_1 \in \mathbb{R}_{\geq 0}.$$

Remark 2 It is worth noticing that one could also define the simplicial volume for non-orientable and non-connected manifolds. Indeed, if M is not orientable, we can simply set $\|M\| := \|\widetilde{M}\|/2$, where $\widetilde{M} \to M$ denotes the orientable double cover of M. Similarly, if M is not connected, then its simplicial volume is just the sum of the simplicial volumes of its connected components.

Computing the exact value of the simplicial volume of a given closed manifold is usually very challenging, and when it is non-zero, there are by now only few cases known: hyperbolic manifolds [Thurston, 1979; Gromov, 1982] (Chapters 2 and 12), manifolds covered by $\mathbb{H}^2 \times \mathbb{H}^2$ [Bucher, 2008], and Hilbert modular surfaces [Löh and Sauer, 2009]. Its non-vanishing is known in more cases: for example, for all closed locally symmetric spaces of non-compact type [Lafont and Schmidt, 2006]; for all closed orientable Riemannian manifolds with sectional curvature bounded from above by a negative number [Gromov, 1982; Thurston, 1979; Inoue and Yano, 1982]; for surface bundles over surfaces [Bucher, 2009]; for complex hyperbolic surfaces [Pieters, 2018b]; and for manifolds satisfying a certain negativity condition on their Ricci curvature [Connell and Wang, 2019, 2020] (Chapter 3). On the other hand, recently Heuer and Löh [2021a] proved that the spectrum of simplicial volume of oriented closed connected n-manifolds is dense in $\mathbb{R}_{\geq 0}$ for all $n \geq 4$ (see Chapter 7).

In many situations, one can still provide estimates of the simplicial volume, which in some cases lead to vanishing results. A classical result is contained in the following:

Proposition 3 *Let $f : M \to N$ be a continuous map between oriented closed connected n-manifolds of degree* $\deg(f)$. *Then, we have*

$$\|M\| \geq |\deg(f)| \cdot \|N\|,$$

with equality whenever f is a covering.

The previous result relating simplicial volume to mapping degree readily shows that manifolds admitting self maps of degree greater than or equal to 2 have vanishing simplicial volume. Hence, we easily deduce that all spheres and tori (or, more generally, products with spheres or tori) of dimension $n \geq 1$ have zero simplicial volume:

$$\|S^n\| = 0 \quad \text{and} \quad \|T^n\| = 0.$$

Moreover, Proposition 3 also implies that the simplicial volume is in fact a homotopy invariant.

Corollary 4 *If M and N are homotopy equivalent, then $\|M\| = \|N\|$.*

Actually, the previous statement can be strengthened as follows: the simplicial volume of essential manifolds (Chapter 4) only depends on their classifying map into a model of their classifying space [Gromov, 1982, Corollary (B), p. 40]. This is a consequence of a deep theorem called *Gromov's mapping theorem* [Gromov, 1982, mapping theorem, p. 40], whose statement and proof are discussed in Chapter 1 (Theorem 1.2).

Simplicial Volume versus Riemannian Geometry

Although simplicial volume could appear at first glance to be only a topological invariant, it actually encodes a lot of information on the geometry carried by the manifold in question. For instance, the simplicial volume of a negatively curved manifold is proportional to its Riemannian volume, thus justifying the name of the invariant. This striking result, which extends, for example, Gauss–Bonnet theorem for hyperbolic manifolds in all dimensions, is known as Gromov and Thurston's *proportionality principle* [Thurston, 1979; Gromov, 1982]. In this collection, we will provide two different proofs: one for hyperbolic manifolds (Chapter 2) and one for non-positively curved ones (Chapter 12). However, it is worth mentioning that the result also holds true without any curvature assumption [Strohm (Löh), 2004; Löh, 2006a; Frigerio, 2011].

Theorem 5 (Proportionality principle [Thurston, 1979; Gromov, 1982]) *Let M be a closed Riemannian manifold. Then the ratio between its simplicial and Riemannian volume only depends on the isometry type of the universal Riemannian covering of M.*

Bounded Cohomology of Spaces and the Duality Principle

For many computations of the simplicial volume, it is more convenient to work with the dual theory, called *bounded cohomology*. Bounded cohomology was first introduced by Johnson [1972] and Trauber in relation to problems on Banach algebras. However, it was only after the pioneering paper by Gromov [1982], which extended the definition from groups to spaces, that it started to spread as an independent and active field [Ivanov, 1987; Noskov, 1991; Monod, 2001; Frigerio, 2017].

We recall here the definition of bounded cohomology of spaces and how it provides information on the simplicial volume via the Duality Principle 10. We refer the reader to Chapter 1 for further results on bounded cohomology of spaces. Let X be a topological space and let $C^n(X; \mathbb{R})$ denote the space of real singular n-cochains on X. We can endow $C^n(X; \mathbb{R})$ with an ℓ^∞-norm as follows:

$$\|f\|_\infty := \sup_{\sigma \in \mathcal{S}_n(X)} |f(\sigma)|,$$

where $\mathcal{S}_n(X)$ denotes the space of all n-singular simplices in X. Then, the previous norm induces a seminorm in cohomology, called ℓ^∞-*seminorm*, as follows:

$$\|\varphi\|_\infty := \inf\{\|f\|_\infty \mid [f] = [\varphi] \in \mathrm{H}_n(X;\mathbb{R})\}.$$

If we consider now the subspace $\mathrm{C}_b^\bullet(X;\mathbb{R}) \subseteq \mathrm{C}^\bullet(X;\mathbb{R})$ of *bounded cochains*, that is, $f \in \mathrm{C}_b^\bullet(X;\mathbb{R})$ if and only if $\|f\|_\infty < +\infty$, we obtain a subcomplex $(\mathrm{C}_b^\bullet(X;\mathbb{R}), \delta^\bullet)$ of the standard singular complex, since the coboundary operator sends bounded cochains to bounded cochains. Hence, we get the following definition:

Definition 6 Let X be a topological space. We define the *nth bounded cohomology group of X* with real coefficients $\mathrm{H}_b^\bullet(X;\mathbb{R})$ to be the homology of the bounded cochain complex $(\mathrm{C}_b^\bullet(X;\mathbb{R}), \delta^\bullet)$. This actually defines a functor from the category of topological spaces to the one of seminormed vector spaces.

Remark 7 In these proceedings, we will often omit the real coefficients from the notation and simply write $\mathrm{H}_b^\bullet(X)$ if there is no ambiguity.

Remark 8 Since bounded cohomology is a homotopy invariant [Gromov, 1982, p. 38], one could extend the previous definition from spaces to groups by simply defining the bounded cohomology of a discrete group G to be the bounded cohomology of any model for its classifying space BG. However, for many algebraic applications, it will be worth introducing a different approach to bounded cohomology of groups, via resolutions (see the section below).

Remark 9 The bounded cohomology groups are naturally endowed with the ℓ^∞-seminorm. Soma [1997b] constructed some examples of bounded cohomology groups for which the seminorm is not a norm in degree 3 (and so the space is not Banach). This was extended to acylindrically hyperbolic groups in Franceschini et al. [2019], using some of the tools that will be presented in Chapter 9. On the other hand, Matsumoto and Morita [1985] showed that the ℓ^∞-seminorm is always a norm in degree 2, and so $\mathrm{H}_b^2(G;\mathbb{R})$ is always a Banach space.

Notice that by construction the inclusion of the bounded cochain complex into the singular one

$$\iota^\bullet \colon \mathrm{C}_b^\bullet(X;\mathbb{R}) \longrightarrow \mathrm{C}^\bullet(X;\mathbb{R})$$

induces a map in cohomology

$$c_X^\bullet \colon \mathrm{H}_b^\bullet(X;\mathbb{R}) \longrightarrow \mathrm{H}^\bullet(X;\mathbb{R})$$

called the *comparison map*. This map detects the gap between the bounded cohomology and the ordinary one. It also allows us to make explicit the duality between bounded cohomology and simplicial volume (Proposition 12.6). To this end, we first recall the definition of *Kronecker pairing*: Given a topological space X, the Kronecker pairing in degree n is the following bilinear map:

$$
\begin{aligned}
\langle \cdot, \cdot \rangle: \quad \mathrm{H}^n(X; \mathbb{R}) \times \mathrm{H}_n(X; \mathbb{R}) &\longrightarrow \mathbb{R} \\
([\varphi], [\alpha]) &\longmapsto \langle [\varphi], [\alpha] \rangle := f(c),
\end{aligned}
\tag{E1}
$$

where $f \in \mathrm{C}^n(X; \mathbb{R})$ is *any* cocycle representing φ and $c \in \mathrm{C}_n(X; \mathbb{R})$ is *any* cycle representing α. Notice that the definition does not depend on the chosen representatives. The fundamental coclass $[M]^*$ in Proposition 10 is defined as the unique element of $\mathrm{H}^n(M; \mathbb{R})$ such that $\langle [M]^*, [M] \rangle = 1$.

This leads to the following result:

Proposition 10 (Duality Principle [Gromov, 1982]) *Let M be an oriented closed connected n-dimensional manifold. Then, the simplicial volume of M is positive if and only if the comparison map c_M^n in degree n is surjective.*

More precisely, we have

$$
\|M\| = \frac{1}{\|[M]^*\|_\infty},
$$

where $[M]^*$ denotes the fundamental coclass of M and we agree that $1/\infty = 0$.

Bounded Cohomology of Groups

As already mentioned above, bounded cohomology of groups was first defined by Johnson [1972] and Trauber in the seventies, and then studied as such, without applications to the simplicial volume, by many mathematicians throughout the years [Ivanov, 1987; Monod, 2001; Burger and Monod, 2002; Burger and Iozzi, 2009; Frigerio, 2017; Ivanov, 2017]. Indeed, bounded cohomology of groups encodes a lot of information, relating to other invariants of groups. For instance, in low degrees, it has connections with other mathematical objects such as stable commutator length (Chapter 7) or quasimorphisms (Chapters 7, 8).

Computing bounded cohomology, as simplicial volume, in general is a very challenging task. For instance, only recently examples of finitely presented groups whose bounded cohomology is fully computed and nontrivial have been constructed [Fournier-Facio et al., 2021, 2022; Monod, 2022]. Nevertheless, even the partial information that has been obtained up to now via the available techniques turns out to have striking applications (see the end of the introduction).

As it will be useful in several chapters of the present collection, we give a brief introduction to bounded cohomology of discrete groups with trivial $\mathbb{R}$ coefficients in what follows. Recall that after the works by Ivanov [1987, 2017], Burger and Monod [2002], and Monod [2001], one can a priori compute bounded cohomology of a group G with $\mathbb{R}$ coefficients just by taking *any* strong resolution of $\mathbb{R}$ via relatively injective modules. However, we prefer to describe here in detail the most common ones, because they will appear later in this volume.

We begin with the so-called *homogeneous resolution*. Let G be a discrete group and let us denote by

$$C^n(G; \mathbb{R}) := \left\{ f : G^{n+1} \longrightarrow \mathbb{R} \right\}$$

the space of degree n cochains on G. Then, we can define the homogeneous coboundary operator as

$$\delta^n : C^n(G; \mathbb{R}) \longrightarrow C^{n+1}(G; \mathbb{R})$$

$$\delta^n(f)(g_0, \ldots, g_{n+1}) := \sum_{i=0}^{n+1} (-1)^i f(g_0, \ldots, \hat{g}_i, \ldots, g_{n+1}),$$

where $\hat{g}_i$ means that we skip the entry g_i. We can then consider the subcomplex of *bounded cochains* $(C_b^\bullet(G; \mathbb{R}), \delta^\bullet) \subset (C^\bullet(G; \mathbb{R}), \delta^\bullet)$, where $f \in C_b^n(G; \mathbb{R})$ if

$$\|f\|_\infty := \sup_{(g_0, \ldots, g_n) \in G^{n+1}} |f(g_0, \ldots, g_n)| < +\infty.$$

Notice that the previous inclusion is well defined, since the homogeneous coboundary operator sends bounded cochains to bounded cochains.

Definition 11 The *homogeneous resolution* of $\mathbb{R}$ is the following exact cocomplex:

$$0 \longrightarrow \mathbb{R} \xrightarrow{\varepsilon} C_b^0(G; \mathbb{R}) \xrightarrow{\delta^0} C_b^1(G; \mathbb{R}) \xrightarrow{\delta^1} C_b^2(G; \mathbb{R}) \xrightarrow{\delta^2} \cdots,$$

where the augmentation map ε is the inclusion of constant functions.

Notice that each space $C_b^n(G; \mathbb{R})$ is naturally endowed with a G-action given by

$$(g \cdot f)(g_0, \ldots, g_n) = f(g^{-1}g_0, \ldots, g^{-1}g_n),$$

where $f \in C_b^n(G; \mathbb{R})$ and $g_0, \ldots, g_n \in G$. Using the previous resolution, we can define the bounded cohomology of G as follows:

Definition 12 The *bounded cohomology* of G with trivial real coefficients $H_b^\bullet(G; \mathbb{R})$ is defined to be the homology of the cocomplex of G-invariants of the homogeneous resolution:

$$0 \longrightarrow C_b^0(G; \mathbb{R})^G \longrightarrow C_b^1(G; \mathbb{R})^G \longrightarrow \cdots ,$$

where $(\cdot)^G$ denotes the submodule of G-invariants.

Remark 13 It is convenient in many situations (Chapters 9, 11) to work with a subresolution of the homogeneous one, called *alternating*. Recall that a cochain $f \in C_b^n(G; \mathbb{R})$ is *alternating* if for every $\sigma \in S_{n+1}$, we have

$$f(g_{\sigma(0)}, \dots, g_{\sigma(n)}) = \text{sign}(\sigma) f(g_0, \dots, g_n),$$

where $\text{sign}(\sigma)$ denotes the signature of σ. We then denote by $C_{b,alt}^n(G; \mathbb{R}) \subset C_b^n(G; \mathbb{R})$ the subspace of bounded alternating n-cochains. This provides the *alternating homogeneous cocomplex*

$$0 \longrightarrow C_{b,alt}^0(G; \mathbb{R}) \longrightarrow C_{b,alt}^1(G; \mathbb{R}) \longrightarrow \cdots .$$

If we restrict to the subcomplex of G-invariants $(C_{b,alt}^\bullet(G; \mathbb{R})^G, \delta^\bullet)$, we can compute the bounded cohomology of G isometrically [Frigerio, 2017, Section 4.10].

Another useful resolution for computing bounded cohomology is the so-called *bar resolution* (or *inhomogeneous resolution*), whose importance is mainly seen in the computation of bounded cohomology in low degrees (Chapter 8). Notice that every G-invariant n-cochain $f \in C_b^n(G; \mathbb{R})^G$ is completely determined by the values that it takes on the set of $(n+1)$-tuples whose first entry is the neutral element e_G. This suggests to define another cochain complex $(\overline{C}^\bullet(G; \mathbb{R}), \delta^\bullet)$, as follows: The *nth cochain group* is defined as the space

$$\overline{C}^n(G; \mathbb{R}) = \{f : G^n \longrightarrow \mathbb{R}\} ,$$

and the *nth coboundary* operator is given by

$$\overline{\delta}^n(f)(g_1, g_2, \dots, g_{n+1}) = f(g_2, \dots, g_{n+1})$$
$$+ \sum_{k=1}^n (-1)^k f(g_1, g_2, \dots, g_k g_{k+1}, \dots, g_{n+1})$$
$$+ (-1)^{n+1} f(g_1, \dots, g_n),$$

where $f \in \overline{C}^n(G; \mathbb{R})$.

Definition 14 The cochain complex $(\overline{C}^\bullet(G; \mathbb{R}), \delta^\bullet)$ is called the *bar cochain complex*; the corresponding augmented complex is called the *bar resolution*.

Remark 15 Note that the difference between the homogeneous resolution and the bar resolution is that the functions in the latter have one less variable, and invariance is no longer required.

The reader is invited to check that $\overline{\delta}^{n+1} \circ \overline{\delta}^n = 0$ and calculate a few examples, such as

$$\overline{\delta}^0(f) = 0,$$

$$\ker(\overline{\delta}^1) = \{\phi \colon G \longrightarrow \mathbb{R} : \phi(gh) = \phi(h) + \phi(g)\} = \mathrm{Hom}(G, \mathbb{R}).$$

One then defines the *nth bounded cochain* group $\overline{C}_b^n(G; \mathbb{R})$ as the subset of $\overline{C}^n(G; \mathbb{R})$ consisting of bounded functions. Since by linearity the coboundary homomorphism sends bounded cochains into bounded cochains, we get a new cochain complex $(\overline{C}_b^\bullet(G; \mathbb{R}), \overline{\delta}^\bullet)$. Notice that by construction we have an isometric isomorphism $(\overline{C}_b^\bullet(G; \mathbb{R}), \overline{\delta}^\bullet) \cong (C_b^\bullet(G; \mathbb{R}), \delta^\bullet)$. This shows that the bounded cohomology of G with trivial real coefficients can be computed via the *bounded bar resolution*.

Bounded Cohomology and Applications

Recall that in the case of bounded cohomology of spaces we have introduced a comparison map that detects the gap between bounded and ordinary cohomology. Similarly, we can define its algebraic analogue as follows: working with the homogeneous resolution, the inclusion of bounded cochains into arbitrary cochains $C_b^\bullet(G) \hookrightarrow C^\bullet(G)$ induces a map

$$c_G^\bullet \colon \mathrm{H}_b^\bullet(G) \longrightarrow \mathrm{H}^\bullet(G)$$

called the *comparison map*. While in the topological setting the comparison map encodes information on the simplicial volume of the manifold in question, in the algebraic setting it yields information about the group G.

Remark 16 Notice that in degree zero the comparison map is always the identity and in degree one it coincides with the zero map. For this reason, we usually restrict our attention to the degrees $n \geq 2$.

We conclude this section by recalling some well-known applications of bounded cohomology of groups. First, we can detect some geometric properties of groups in terms of the behaviour of their comparison map. As a most striking example, we cite Mineyev's theorem, which invokes the bounded cohomology theory *with coefficients* (see Monod [2001]; Frigerio [2017]):

Theorem 17 [Mineyev, 2002, theorem 3] *For a finitely presentable group G, the following statements are equivalent:*

- G *is hyperbolic.*
- *The comparison map* $H_b^2(G, V) \to H^2(G, V)$ *is surjective for any bounded G-module V.*
- *The comparison map* $H_b^n(G, V) \to H^n(G, V)$ *is surjective for any $n \geq 2$ and any bounded G-module V.*

From a topological point of view, via the Duality Principle 12.6, the previous result readily implies that the simplicial volume of a closed negatively curved manifold is non-zero.

On the other side of the spectrum, we have the older result of Trauber and Johnson:

Theorem 18 [Johnson, 1972] *If the group G is amenable, then $H_b^n(G; \mathbb{R}) = 0$ for all $n \geq 1$.*

In contrast to Mineyev's result, the previous theorem, together with the mapping theorem (Theorem 1.2), imply the vanishing of the simplicial volume of every closed connected manifold of positive dimension with amenable fundamental group.

As shown by the previous two results, bounded cohomology can behave very differently from the ordinary one. Indeed, for $k \geq 1$, the bounded cohomology of $\mathbb{Z}^k$ vanishes in positive degree by Theorem 18, while its usual cohomology does not. On the other hand, *non-abelian* free groups F_k have $H^n(F_k; \mathbb{R}) = 0$ for all $n \geq 2$, while both the second and the third bounded cohomology groups $H_b^2(F_k; \mathbb{R})$, $H_b^3(F_k; \mathbb{R})$ are infinite-dimensional [Brooks, 1981; Mitsumatsu, 1984; Soma, 1997a].

However, for certain classes of groups, it is believed that there is no difference between bounded and ordinary cohomology. The comparison map is the main character of the following well-known conjecture [Dupont, 1979; Monod, 2006a], which is still open for many groups (for more details, see Chapter 10 and Conjecture 10.15 therein):

Conjecture 19 [Dupont, 1979; Monod, 2006a] If G is a connected semisimple Lie group without compact factors and with finite center, then the comparison map of G is an isomorphism.

Finally, bounded cohomology is a fundamental tool in rigidity theory. We mention here three instances. The first one exploits the fact that a surjective homomorphism $G \to H$ induces an embedding in bounded cohomology in degree 2 [Huber, 2013, Theorem 2.14]. This implies that if the bounded cohomology of G is small (say, it vanishes or is finite-dimensional) and the one of H is large (say, it is infinite-dimensional), then G cannot surject onto H. This idea is exploited for instance in the proof by Bestvina and Fujiwara [2002] of a superrigidity theorem for mapping class groups, initially proven by Farb and Masur [1998].

Another example that showcases the importance of bounded cohomology in ridigity theory is the following: Group actions on certain geometric objects produce bounded cohomology classes, whose vanishing or non-vanishing captures properties of the actions. For instance, for higher rank lattices, one deduces that all such actions are elementary in a suitable sense [Burger and Monod, 1999]. Geometric objects to which this philosophy applies include $\mathbb{H}^n$ [Gromov, 1993, 7.E1; Sela, 1992], the circle [Ghys, 2001] (see also Chapter 11), a large class of negatively curved spaces [Monod, 2006b; Monod and Shalom, 2004, 2006; Mineyev et al., 2004], and finite-dimensional CAT(0) cube complexes [Chatterji et al., 2016].

Finally, bounded cohomology comes into play when studying representations of discrete groups into Lie groups, leading to the study of *maximal representations*: we refer the reader to Chapter 11 for a detailed account.

Part I

SIMPLICIAL VOLUME

1

Gromov's Mapping Theorem via Multicomplexes

Marco Moraschini[*]

The aim of this short note is to give an introduction to the theory of *multicomplexes* and explain its role in the proof that bounded cohomology of topological spaces only depends on their fundamental groups:

Theorem 1.1 [Gromov, 1982; Ivanov, 2017; Frigerio and Moraschini, 2019]
Let $f: X \to Y$ be a continuous map between path-connected CW-complexes such that $\pi_1(f): \pi_1(X) \to \pi_1(Y)$ is an isomorphism. Then, for every $n \geq 0$, the induced map

$$\mathrm{H}_b^n(f): \mathrm{H}_b^n(Y) \longrightarrow \mathrm{H}_b^n(X)$$

is an isometric isomorphism.

Notice that Theorem 1.1 can be extended in two directions. On the one hand, we can consider all topological spaces, and on the other, it is sufficient to require that $\pi_1(f)$ is an epimorphism with amenable kernel. This latter general formulation is usually known as *Gromov's Mapping Theorem*:

Theorem 1.2 [Gromov, 1982; Ivanov, 2017; Frigerio and Moraschini, 2019]
Let X and Y be path-connected topological spaces and let $f: X \to Y$ be a continuous map such that $\pi_1(f): \pi_1(X) \to \pi_1(Y)$ is an epimorphism with amenable kernel. Then, for every $n \geq 0$, the induced map

$$\mathrm{H}_b^n(f): \mathrm{H}_b^n(Y) \longrightarrow \mathrm{H}_b^n(X)$$

is an isometric isomorphism.

Unfortunately, Gromov's Mapping Theorem is not a straightforward corollary of Theorem 1.1 and so we prefer to omit here all the technical details

---

[*] Supported by CRC 1085, *Higher Invariants* (Universität Regensburg, funded by the DFG).

3

for deducing it from Theorem 1.1. However, once readers are familiar with all the ideas described in the sequel, they will be more than encouraged to check the missing steps in order to prove Gromov's Mapping Theorem. We refer the reader to Frigerio and Moraschini [2019, theorem 3] to see how to extend Theorem 1.1 from path-connected CW-complexes to all path-connected topological spaces (the proof makes essential use of the invariance of bounded cohomology under weak homotopy equivalences proved by Ivanov [Ivanov, 2017]) and to Frigerio and Moraschini [2019, chapter 5] for the complete proof of Gromov's Mapping Theorem

1.1 Strategy of the Proof of Theorem 1.1

Before introducing the notion of multicomplexes and their properties, we outline here the strategy for proving Theorem 1.1.

First recall that a path-connected CW-complex X is said to be *aspherical* if $\pi_n(X) = 0$ for all $n \geq 2$. Notice that the homotopy type of aspherical CW-complexes is completely characterized by their fundamental group [Hatcher, 2002, theorem 1B.8], that is, they are homotopy equivalent if and only if they have an isomorphic fundamental group. For this reason, aspherical spaces are usually represented by the symbol $K(G, 1)$, where G denotes their fundamental group. Moreover, for every path-connected CW-complex X, there always exists a unique (up to homotopy) map

$$f_X : X \longrightarrow K(\pi_1(X), 1),$$

called the *classifying map*, such that $\pi_1(f_X)$ is an isomorphism of fundamental groups [Hatcher, 2002, proposition 1B.9].

We claim now that Theorem 1.1 is a straightforward consequence of the following:

Theorem 1.3 [Gromov, 1982 corollary (A), p. 40; Frigerio and Moraschini, 2019] *Let X be a path-connected CW-complex. Then, there exists an aspherical space $K(\pi_1(X), 1)$ and a classifying map $f_X : X \to K(\pi_1(X), 1)$ such that for every $n \geq 0$, the induced map*

$$H_b^n(f_X) : H_b^n(K(\pi_1(X), 1)) \longrightarrow H_b^n(X)$$

is an isometric isomorphism.

Remark 1.4 Since classifying maps (and aspherical spaces) are unique up to homotopy, it is immediately apparent that the previous statement holds in fact for *any* space $K(\pi_1(X), 1)$ and *any* classifying map $f_X : X \to K(\pi_1(X), 1)$.

However, since in the proof we will provide an explicit classifying map, we prefer to state the theorem as above.

Assuming Theorem 1.3, we are now ready to prove Theorem 1.1:

Proof of Theorem 1.1 Let $f: X \to Y$ be a continuous map between path-connected CW-complexes such that $\pi_1(f): \pi_1(X) \to \pi_1(Y)$ is an isomorphism. Suppose that

$$f_X: X \longrightarrow K(\pi_1(X), 1)$$

and

$$f_Y: Y \longrightarrow K(\pi_1(Y), 1)$$

are the classifying maps provided by Theorem 1.3. Then, via the general theory of aspherical spaces [Hatcher, 2002, proposition 1B.9], we know that there exists a continuous map

$$g: K(\pi_1(X), 1) \longrightarrow K(\pi_1(Y), 1),$$

induced by the following group homomorphism:

$$\pi_1(K(\pi_1(X), 1)) \xrightarrow[\cong]{\pi_1(f_X)^{-1}} \pi_1(X) \xrightarrow{\pi_1(f)} \pi_1(Y) \xrightarrow[\cong]{\pi_1(f_Y)} K(\pi_1(Y), 1).$$

Notice that, by construction and by the asphericity of $K(\pi_1(Y), 1)$, the map g makes the following diagram

$$
\begin{array}{ccc}
X & \xrightarrow{\quad f \quad} & Y \\
{\scriptstyle f_X}\downarrow & & \downarrow{\scriptstyle f_Y} \\
K(\pi_1(X), 1) & \xrightarrow{\quad g \quad} & K(\pi_1(Y), 1)
\end{array}
$$

commute up to homotopy (see again [Hatcher, 2002, proposition 1B.9]). Moreover, g is a weak homotopy equivalence because it induces isomorphisms on all homotopy groups. By Whitehead theorem [Hatcher, 2002, theorem 4.5], this shows that g is in fact a homotopy equivalence. Since bounded cohomology is a homotopy invariant, we then get the following commutative diagram

$$
\begin{array}{ccc}
H_b^n(K(\pi_1(Y), 1)) & \xrightarrow{\;H_b^n(g)\;} & H_b^n(K(\pi_1(X), 1)) \\
{\scriptstyle H_b^n(f_Y)}\downarrow & & \downarrow{\scriptstyle H_b^n(f_X)} \\
H_b^n(Y) & \xrightarrow[\;H_b^n(f)\;]{} & H_b^n(X)
\end{array}
$$

in every degree $n \geq 0$, where the vertical arrows are isometric isomorphisms by Theorem 1.3 and the lower horizontal arrow is an isometric isomorphism because g is a homotopy equivalence. By the 2-out-3 property of isometric isomorphisms, we conclude that also $H_b^n(f)$ is an isometric isomorphism for all $n \geq 0$. This finishes the proof.   □

We spend the remaining part of this note outlining the proof of Theorem 1.3.

1.2 Multicomplexes

When we study simplicial structures in topology, we can order the most common simplicial structures according to the amount of degeneracies allowed in their simplices. In this spectrum, one can consider simplicial complexes [Frigerio and Moraschini, 2019, definition 1.1.2; Hatcher, 2002] as the simplicial structure with the least number of degeneracies and simplicial sets [May, 1992] as the simplicial structure admitting the largest number of *degenerated* objects. Then Eilenberg and Zilber [1950] introduced another simplicial structure called *semi-simplicial sets* (now known as Δ-*sets* or Δ-*complexes*), which differ from simplicial sets because of the lack of degeneracies. Hence, we can consider Δ-complexes as an intermediate simplicial structure lying in the middle of the spectrum between simplicial complexes and simplicial sets.

According with this terminology, one could simply define a *multicomplex* to be an *unordered* Δ-complex in which every simplex has distinct vertices or, equivalently, as a *symmetric* simplicial set in which every non-degenerate simplex has distinct vertices [Frigerio and Moraschini, 2019, proposition 1.3.1]. We refer the reader to the literature for the notion of symmetric simplicial sets [Grandis, 2001a,b].

After this brief introduction to the comparison between multicomplexes and the other well-known simplicial structures in topology, we are ready to provide a formal definition. We begin by stating the precise and concise definition that Gromov himself gave of a multicomplex [Gromov, 1982]: a *multicomplex* is "a set K divided into the union of closed affine simplices Δ_i, $i \in I$, such that the intersection of any two simplices $\Delta_i \cap \Delta_j$ is a (simplicial) subcomplex in Δ_i as well as in Δ_j." Despite this definition already containing all the information we need for deducing all the properties of a multicomplex, we provide here a definition closer to the more modern abstract-algebraic approach to simplicial structures:

Definition 1.5 [Frigerio and Moraschini, 2019, definition 1.1.1] Let $\mathcal{P}_f(V)$ denote the set of finite subsets of a given set V. A *multicomplex* K is a triple

$$K = \left(V, I = \bigcup_{A \in \mathcal{P}_f(V)} I_A, \Omega \right),$$

where

1. V is any set and it is called the *set of vertices* of K;
2. for every $A \in \mathcal{P}_f(V)$, I_A is a (possibly empty) set that denotes the *set of simplices with vertex set A*;
3. if $A = \{v\}$ is a singleton, then I_A is also a singleton;
4. Ω is a set of maps $\{\partial_{A,B} \colon I_A \to I_B, A, B \in \mathcal{P}_f(V), A \supseteq B\}$ (called *boundary maps* of the multicomplex) such that $\partial_{A,A} = \mathrm{Id}_A$ for every $A \in \mathcal{P}_f(V)$, and $\partial_{B,C} \circ \partial_{A,B} = \partial_{A,C}$ if $A \supseteq B \supseteq C$.

Remark 1.6 Notice that the absence of degenerated simplices in the definition of multicomplexes makes the study of their combinatorics much easier than that of simplicial sets. Moreover, it also has strong implications for *simplicial maps*, which now cannot shrink simplices into simplices of lower dimension [Frigerio and Moraschini, 2019, remark 1.1.4].

Remark 1.7 We emphasize here that our definition does not provide a topological space but only an algebraic datum. Nevertheless, associated to each multicomplex K there exists a canonical topological space called *geometric realization*, as denoted by $|K|$, obtained by endowing K with the weak topology associated to its decomposition into simplices. In fact, following Hatcher's notation [Hatcher, 2002, page 533] (see also [Fritsch and Piccinini, 1990]), one can interpret the geometric realization of a multicomplex as a *regular unordered* Δ-complex [Frigerio and Moraschini, 2019, page 24], where regular means *with embedded closed cells*. This latter property reflects the fact that inside a multicomplex simplices must have distinct vertices. We refer the reader to Frigerio and Moraschini [2019, section 1.2] for a formal definition of geometric realization.

Example 1.8 [Moraschini, 2018, example 1.3.2]   As mentioned above, we can distinguish different simplicial structures just by looking at the amount of degenerated simplices. This can be rephrased in terms of computing the *minimal* number of vertices that we need to construct a simplicial object whose geometric realization is homeomorphic to a given topological space. Suppose, for instance, that we want to consider a simplicial complex, a simplicial set, and a multicomplex with the minimal number of vertices such that their geometric realization is homeomorphic to the cone over the circle in such a way that the apex of the cone p lies in the set of vertices. Then we have the following cases:

1. The minimal simplicial complex K has four vertices. Indeed, if $V = \{p, x_1, x_2, x_3\}$ denotes the vertex set, K consists in the union of three 2-simplices, whose vertices are the subsets $\{p, x_i, x_j\} \subset V$, for $1 \leq i < j \leq 3$. Of course, K is also a multicomplex and a Δ-complex.

2. The minimal multicomplex Z has three vertices. Indeed, Z can be constructed by gluing two 2-simplices together along all but one of their facets. One can check that Z is no longer a simplicial complex, but it is still a Δ-complex.

3. The minimal Δ-complex T has only two vertices. Indeed, let $\{x_1, x_2, x_3\}$ be a labeling of the vertices of a 2-simplex, where $x_1 = p$, and let us call e_{ij} the edge between the vertices x_i and x_j, for $1 \leq i < j \leq 3$. Then we glue the edge e_{12} with the edge e_{13} via an affine map, sending x_1 to x_1 and x_2 to x_3. The resulting complex provides a Δ-complex realizing the desired cone. Since K has only 2 vertices, K does not contain embedded simplices. This shows that K is neither a multicomplex nor a simplicial complex.

1.3 The Singular Multicomplex

Recall that we introduced multicomplexes in order to prove Theorem 1.3. The first step in our construction consists in associating to each path-connected CW-complex X a multicomplex $\mathcal{K}(X)$ whose geometric realization has the same homotopy type of X. More precisely, we will construct the so-called *singular multicomplex* $\mathcal{K}(X)$, which plays the same role in the theory of multicomplexes as the well-known *singular set* $\mathcal{S}(X)$ in the one of Δ-complexes and simplicial sets [Milnor, 1957; Goerss and Jardine, 1999]. However, since Δ-complexes and simplicial sets are not in general multicomplexes, we have to refine the standard construction. We will see that the main differences between $\mathcal{K}(X)$ and $\mathcal{S}(X)$ consist in the following facts:

1. Only singular simplices which are *injective* on the vertices of $|\Delta^n|$ are allowed in the construction of $\mathcal{K}(X)$.

2. The geometric simplices of $|\mathcal{K}(X)|$ are not endowed with a preferred orientation of their vertices.

We are now ready to give a precise definition of $\mathcal{K}(X)$:

Definition 1.9 [Frigerio and Moraschini, 2019, chapter 2] Let X be a path-connected CW-complex and let us denote by $\mathcal{S}_n(X)$ the set of singular n-simplices $\sigma : |\Delta^n| \to X$ and by $\mathcal{K}_n(X) \subset \mathcal{S}_n(X)$ the subset of those singular n-simplices which are injective on the vertices of $|\Delta^n|$. Then we say that two singular n-simplices $\sigma, \sigma' : |\Delta^n| \to X$ are *equivalent* if they differ by a precomposition via an affine diffeomorphism of $|\Delta^n|$, that is, $\sigma \circ \tau = \sigma'$ for some affine diffeomorphism $\tau : |\Delta^n| \to |\Delta^n|$. We denote by $\overline{\mathcal{K}}_n(X)$ the set of equivalence classes of simplices in $\mathcal{K}_n(X)$.

The *singular multicomplex* $\mathcal{K}(X)$ associated to X is given by the triple

$$\mathcal{K}(X) = (V, I, \Omega),$$

where V consists of the points of X; for every $A \subseteq V$ of cardinality $(n + 1)$, we set I_A to be the set of all equivalence classes of singular n-simplices $[\sigma]$ in $\overline{\mathcal{K}}_n(X)$ such that $\sigma(|\Delta^n|^0) = A$, that is, any representative of $[\sigma]$ maps the vertices of the standard n-simplex onto A; we define Ω as follows: if $[\sigma] \in I_A$, then for every $B \subseteq A$, we set $\partial_{A,B}[\sigma]$ to be the equivalence class of the unique face of σ which maps its vertices onto B.

Remark 1.10 Notice that as soon as X contains uncountably many points, the associated singular multicomplex $\mathcal{K}(X)$ has uncountably many vertices. Then, if X is a CW-complex of positive dimension (i.e., X is not a set of vertices), there exist uncountably many singular simplices (in every dimension) that are injective on the set of vertices. This shows that $\mathcal{K}(X)$ contains uncountably many simplices in every dimension. In order to make the study of the combinatorics of $\mathcal{K}(X)$ more accessible, we will explain in the following sections how to *reduce* the size of $\mathcal{K}(X)$ without changing its homotopy type.

Remark 1.11 Every point in $|\mathcal{K}(X)|$ can be easily described in terms of barycentric coordinates. Indeed, every n-cell E of $|\mathcal{K}(X)|$ with set of vertices $\{x_0, \cdots, x_n\}$ corresponds to an equivalence class $[\sigma_E] \in \overline{\mathcal{K}}_n(X)$, where σ_E is an n-singular simplex sending each vertex $e_i \in |\Delta^n|$ to $x_i \in X$. Hence, we can parametrize the points lying in E simply as

$$([\sigma_E], t_0 x_0, \ldots, t_n x_n),$$

where $\sum_{i=0}^{n} t_i = 1$ and $t_i \geq 0$ for every $i = 0, \cdots, n$. Notice that this description is almost unique [Frigerio and Moraschini, 2019, remark 1.2.1].

As we anticipated above, the most important feature of $\mathcal{K}(X)$ is that it preserves the homotopy type of X. This can be proved by considering the natural projection

$$S_X : |\mathcal{K}(X)| \longrightarrow X \tag{1.1}$$

defined as follows:

$$S_X(([\sigma], t_0 x_0, \ldots, t_n x_n)) = \sigma(t_0 e_0 + \cdots + t_n e_n),$$

where we kept the notation of Remark 1.11. By construction S_X is well-defined and continuous.

Theorem 1.12 [Gromov, 1982, example (c), p. 42; Frigerio and Moraschini, 2019, corollary 2.1.2] *Let X be a path-connected CW-complex. Then, the natural projection* (1.1)

$$S_X \colon |\mathcal{K}(X)| \longrightarrow X$$

is a homotopy equivalence.

Remark 1.13 A natural question is whether the previous statement holds for *any* topological space, as in the case of the singular sets in the theory of Δ-complexes and simplicial sets [Milnor, 1957]. First, recall that for arbitrary topological spaces, the most natural corresponding statement would be to prove that the canonical projection S_X is a *weak homotopy equivalence*, that is, it induces isomorphisms on all homotopy groups. As far as we know, our techniques do not lead to a statement for *any* topological space. However, the previous theorem is in fact a particular case of a more general statement for a larger family of spaces [Frigerio and Moraschini, 2019, theorem 1].

Corollary 1.14 *Let X be a path-connected CW-complex and let*

$$S_X \colon |\mathcal{K}(X)| \longrightarrow X$$

be the natural projection (1.1). *Then, for every $n \geq 0$, the induced map*

$$H_b^n(S_X) \colon H_b^n(X) \longrightarrow H_b^n(|\mathcal{K}(X)|)$$

is an isometric isomorphism.

Proof It is sufficient to apply Theorem 1.12 and the fact that bounded cohomology is a homotopy invariant.   □

1.4 Complete Multicomplexes

Recall that our final goal is to construct for any given path-connected CW-complex X a classifying map f_X inducing isometric isomorphisms on all bounded cohomology groups. To this end, it is convenient to better understand the homotopy of multicomplexes. The first step in this direction consists in the study of *complete* multicomplexes, which correspond to *Kan complexes* in the theory of simplicial sets. (We refer the reader to Frigerio and Moraschini [2019, remark 3.12] for a detailed comparison between these two notions.)

Definition 1.15 [Frigerio and Moraschini, 2019, definition 3.1.1] We say that a continuous map $f: |\Delta^n| \to |K|$ is a ∂-*embedding* if its restriction to the boundary $f|_{\partial|\Delta^n|}: \partial|\Delta^n| \to |K|$ is a simplicial embedding (i.e., an injective simplicial map).

A multicomplex K is said to be *complete* if every ∂-embedding map $f: |\Delta^n| \to |K|$ is homotopic relative to $\partial|\Delta^n|$ to a simplical embedding $\iota: |\Delta^n| \to |K|$.

Example 1.16 [Frigerio and Moraschini, 2019, example 3.1.3] We list here some easy examples that could help the reader to better understand the combinatorics of complete multicomplexes:

1. If K is a connected complete one-dimensional multicomplex, then K must be a segment.
2. If K is a connected complete multicomplex with infinite vertex set V, then K is infinite-dimensional.
3. If a simplicial complex K is a connected complete multicomplex, then K is the full simplicial complex on K^0, that is, the set of vertices of K.

Special Spheres

The main reason we are interested in complete multicomplexes is that we can describe their homotopy groups in a simplicial way. To this end, we have to introduce the notion of *special spheres* (see Figure 1.1).

Definition 1.17 [Frigerio and Moraschini, 2019, definition 3.2.4] We say that two n-simplices Δ_1 and Δ_2 of a multicomplex K are *compatible* if $|\partial\Delta_1| = |\partial\Delta_2|$ in the geometric realization of K. Moreover, given an n-simplex $\Delta \subset K$, we denote by $\pi(\Delta)$ the set of all n-simplices of K that are compatible with Δ.

Remark 1.18 Notice that the multicomplex generated by the union of two compatible n-simplices Δ_n and Δ_s is naturally homeomorphic to an n-sphere

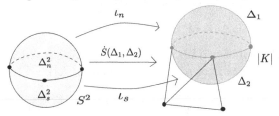

Figure 1.1 A special sphere

(see Figure 1.1). For this reason, a multicomplex generated by two compatible n-simplices will be denoted by $\dot{S}^n$.

Definition 1.19 Let $\dot{S}^n$ be an n-dimensional multicomplex obtained by gluing together two compatible n-simplices Δ_n and Δ_s. Let K be a multicomplex, and let $\Delta_1, \Delta_2 \subset K$ be two compatible n-simplices. We say that a continuous map from the n-sphere to $|K|$

$$\dot{S}^n(\Delta_1, \Delta_2)\colon S^n \cong |\dot{S}^n| \longrightarrow |K|$$

is a *special sphere* (see Figure 1.1) if the restrictions

$$\iota_n := \dot{S}^n(\Delta_1, \Delta_2)|_{|\Delta_n|}\colon |\Delta_n| \longrightarrow |\Delta_1| \subset |K|$$

and

$$\iota_s := \dot{S}^n(\Delta_1, \Delta_2)|_{|\Delta_s|}\colon |\Delta_s| \longrightarrow |\Delta_2| \subset |K|$$

are simplicial embeddings.

Using special spheres and Theorem 1.12, one can show that given any path-connected CW-complex X, the associated singular multicomplex $\mathcal{K}(X)$ is complete:

Theorem 1.20 [Gromov, 1982, example (a), p. 42; Frigerio and Moraschini, 2019, theorem 3.2.3] *Let X be a path-connected CW-complex. Then the associated singular multicomplex $\mathcal{K}(X)$ is complete.*

Homotopy Groups of Complete Multicomplexes

Special spheres are also useful in the study of homotopy groups of complete multicomplexes. More precisely, we begin with the following definition:

Definition 1.21 [Frigerio and Moraschini, 2019, definition 3.2.4] Let K be a multicomplex and let $\Delta_1, \Delta_2 \subset K$ be two compatible n-simplices. We say that Δ_1 and Δ_2 are *homotopic* if the corresponding (pointed) special sphere

$$\dot{S}^n(\Delta_1, \Delta_2)\colon (S^n, s_0) \cong (|\dot{S}^n|, s_0) \longrightarrow (|K|, x_0)$$

is the trivial element of $\pi_n(|K|, x_0)$, where s_0 is a chosen vertex of $\dot{S}^n$ which is mapped to the vertex x_0 of Δ_1 (or, equivalently, of Δ_2).

Using the notion of homotopic simplices, we are now ready to describe the homotopy groups of complete multicomplexes in a simplicial/combinatorial way (compare with Goerss and Jardine [1999, section I.11 p. 60] and May [1992, definition 3.6] for the case of simplicial sets).

Theorem 1.22 [Frigerio and Moraschini, 2019, theorem 3.2.5] *Let K be a complete multicomplex and let $\Delta_0 \subset K$ be an n-simplex of K. Suppose that x_0 is a fixed vertex of Δ_0. Then the map*

$$\Theta \colon \pi(\Delta_0) \longrightarrow \pi_n(|K|, x_0)$$
$$\Delta_1 \longmapsto \left[\dot{S}^n(\Delta_0, \Delta_1)\right]$$

is surjective and $\Theta(\Delta_1) = \Theta(\Delta_2)$ if and only if Δ_1 and Δ_2 are homotopic simplices of K.

Remark 1.23 Notice that the previous result allows us to give an explicit description of the homotopy type of any given path-connected CW-complex X. Indeed, it is sufficient to replace X with $\mathcal{K}(X)$ (Theorem 1.12) and then apply the previous result.

However, the theorem above is far from being optimal. Indeed, it shows that complete multicomplexes contain too many redundant simplices (i.e., the homotopic ones). For this reason, we will introduce *minimal* multicomplexes in the next section.

1.5 Minimal Multicomplexes

Our goal in this section is to explain how to "reduce" the size of complete multicomplexes without affecting their homotopy type. We begin with the following definition:

Definition 1.24 [Frigerio and Moraschini, 2019, lemma 3.4.2] A multicomplex K is called *minimal* if it does not contain any pair of homotopic simplices.

Example 1.25 [Moraschini, 2018, example 3.4.4] The following examples show that a priori there are no implications between the notions of completeness and minimality:

1. Let $\dot{S}^1$ be the multicomplex obtained by gluing two 1-simplices along their common boundaries. Then, $\dot{S}^1$ is minimal but *not* complete (since the fundamental group of S^1 is nontrivial).
2. Similarly, if K is the multicomplex described in Example 1.8 (2), then K is complete but *not* minimal.

Despite the previous examples showing that completeness does not imply minimality (or vice versa), we get the following easy corollary of Theorem 1.22 (compare with Goerss and Jardine [1999, section I.11, p. 60] for a similar result for simplicial sets):

Theorem 1.26 [Frigerio and Moraschini, 2019, theorem 3.4.5]   *Let K be a complete and minimal multicomplex and let $\Delta_0 \subset K$ be an n-simplex of K. Suppose that x_0 is a fixed vertex of Δ_0. Then the map*

$$\Theta : \pi(\Delta_0) \longrightarrow \pi_n(|K|, x_0)$$
$$\Delta_1 \longmapsto \left[\dot{S}^n(\Delta_0, \Delta_1)\right]$$

is a bijection.

The previous theorem shows that in order to explicitly describe homotopy groups of a multicomplex, we need to consider multicomplexes that are both complete and minimal. To this end, we have to produce a standard construction that associates to each complete multicomplex K a complete and minimal one L with the same homotopy type. The procedure is explained in the following easy example:

Example 1.27   Let us consider the situation of Example 1.25 (2). We know that K is a complete (but not minimal!) multicomplex such that $|K|$ is contractible. We want to construct a complete and minimal multicomplex L such that $L \subseteq K$ and the inclusion map is a homotopy equivalence.

The construction goes as follows: The set of vertices of L coincides with the set of vertices of K. Then among the sets of edges with the same endpoints, we arbitrary choose a unique 1-simplex for each homotopy class of 1-simplices in K. This produces a 1-dimensional multicomplex L^1 with 3 vertices and a unique 1-simplex between each pair of vertices. Then, again we choose a unique 2-simplex among each homotopy class of 2-simplices of K whose boundary lies in L^1. It is immediate to see that in this situation L is nothing more than a standard 2-simplex inside K.

Notice that the inclusion $i : |L| \to |K|$ is a homotopy equivalence and that L is both complete and minimal as desired.

In fact, the previous construction works in general, as described by the following theorem (compare with May [1992, thoerems 9.5 and 9.8] for similar results in the theory of simplicial sets):

Theorem 1.28 [Frigerio and Moraschini, 2019, theorem 3.4.6]   *Let K be a complete multicomplex. Then, there exists a subcomplex $L \subset K$ such that*

1. *L is complete and minimal;*
2. *L has the same vertex set as K;*
3. *The inclusion i : |L| → |K| is a homotopy equivalence.*

Remark 1.29 Notice that the multicomplex L constructed by the previous theorem is *unique* up to simplicial isomorphisms [Frigerio and Moraschini, 2019, theorem 3.4.6]. For this reason, we will often refer to it as the minimal and complete multicomplex inside K.

Definition 1.30 Let X be a path-connected CW-complex and let $\mathcal{K}(X)$ be the singular multicomplex associated to it. Then, we will denote by $\mathcal{L}(X)$ the minimal and complete multicomplex contained in $\mathcal{K}(X)$.

Using the previous theorem, we get the following result:

Corollary 1.31 [Frigerio and Moraschini, 2019, corollary 3.4.10] *Let X be a path-connected CW-complex. Then, there exists a complete and minimal multicomplex $\mathcal{L}(X)$ such that*

$$S_X \circ i \colon |\mathcal{L}(X)| \longrightarrow X$$

is a homotopy equivalence, where $i \colon |\mathcal{L}(X)| \to |\mathcal{K}(X)|$ denotes the inclusion and S_X the natural projection (1.1).
In particular, for every $n \geq 0$, the induced map

$$\mathrm{H}_b^n(S_X \circ i) \colon \mathrm{H}_b^n(X) \longrightarrow \mathrm{H}_b^n(|\mathcal{L}(X)|)$$

is an isometric isomorphism.

Remark 1.32 Recall that our goal is to prove Theorem 1.3. To this end, we have to construct a classifying map for X inducing isometric isomorphisms on bounded cohomology groups. Hence, it is convenient to consider $j_X \colon X \to |\mathcal{L}(X)|$ to be the homotopy inverse of $S_X \circ i$. We will show in the next section how to map $|\mathcal{L}(X)|$ into an aspherical space whose fundamental group is isomorphic to $\pi_1(X)$.

1.6 Aspherical Multicomplexes

Let X be a path-connected CW-complex. Our goal is now to explain how to associate to $\mathcal{L}(X)$ an aspherical multicomplex $\mathcal{A}(X)$ in such a way that the bounded cohomologies of their geometric realizations are isometrically isomorphic. To this end, we will describe $\mathcal{A}(X)$ as a (simplicial) quotient of $\mathcal{L}(X)$. We begin with the following definition:

Definition 1.33 [Frigerio and Moraschini, 2019, definition 4.3.1] Let X be a path-connected CW-complex and let $\mathcal{L}(X)$ be the minimal and complete multicomplex associated with it. We denote by Γ the group of all simplicial automorphisms of $\mathcal{L}(X)$ that are homotopic to the identity relative to the 0-skeleton. Moreover, for every $i \geq 1$, we consider the following groups:

$$\Gamma_i := \{g \in \Gamma \mid g|_{K^i} = \mathrm{Id}_{K^i}\},$$

where K^i denotes the i-skeleton of K (i.e., the set of all its i-simplices).

Remark 1.34 Notice that our definition is slightly different from Gromov's original one. We refer the reader to Frigerio and Moraschini [2019, remark 4.3.2] for a detailed discussion of this issue.

We are now interested in understanding the combinatorics of the quotient of $\mathcal{L}(X)$ under the action of the groups Γ_i:

Proposition 1.35 [Frigerio and Moraschini, 2019, proposition 4.3.5] *Let $n \geq i + 1 \geq 2$ be two integers. Let Δ and Δ' be two n-simplices of $\mathcal{L}(X)$. Then, Δ and Δ' lie in the same Γ_i-orbit if and only if they share the same i-skeleton.*

Using the previous result, one can prove that the quotient of $\mathcal{L}(X)$ under the action by Γ_1 is an aspherical complete and minimal multicomplex:

Theorem 1.36 [Frigerio and Moraschini, 2019, corollary 4.3.6] *Let X be a path-connected CW-complex, and let $\mathcal{L}(X)$ be the minimal and complete multicomplex associated with it. Then, the quotient*

$$\mathcal{A}(X) := \mathcal{L}(X)/\Gamma_1$$

is a complete and minimal multicomplex such that $|\mathcal{A}(X)|$ is an aspherical $K(\pi_1(X), 1)$-space.

Since the proof is rather involved, we only give here a rough idea about why the multicomplex $\mathcal{A}(X)$ is expected to be aspherical *once* you already know that $\mathcal{A}(X)$ is both complete and minimal. (We refer the reader to Frigerio and Moraschini [2019, sections 3.5 and 4.3] for a complete proof.)

Idea of the proof under the additional assumptions above By definition of Γ_1, the 1-skeleton of $\mathcal{L}(X)$ is preserved. Hence, the fact that the projection $\pi \colon |\mathcal{L}(X)| \to |\mathcal{A}(X)|$ induces an isomorphism on fundamental groups is an easy consequence of the following two facts: The 1-skeleta of $\mathcal{A}(X)$ and $\mathcal{L}(X)$ are the same and every triangular simplicial loop in $\mathcal{A}(X)^1 = \mathcal{L}(X)^1$ bounds a 2-simplex in $\mathcal{A}(X)$ if and only if it does in $\mathcal{L}(X)$ [Frigerio and Moraschini, 2019, proof of theorem 3.5.3].

Moreover, *if we assume* that $\mathcal{A}(X)$ is complete and minimal, then we can describe the higher homotopy groups of $|\mathcal{A}(X)|$ as in Theorem 1.26. More precisely, the higher homotopy groups of $|\mathcal{A}(X)|$ are detected by compatible simplices in dimension $n \geq 2$. However, Proposition 1.35 shows that in the quotient $\mathcal{A}(X)$ there are no compatible simplices in dimension $n \geq 2$ [Frigerio and Moraschini, 2019, Corollary 4.3.6]. This shows that all higher homotopy groups of $|\mathcal{A}(X)|$ are trivial; hence the thesis. $\square$

Remark 1.37 Despite the previous proof providing just an intuition about the role played by minimal and complete multicomplexes in this step, it is important to mention here that as far as we know we cannot expect an equivalent result by using simplicial sets. We refer the reader to Frigerio and Moraschini [2019, remark 4.3.7] for a thorough discussion of the reason why multicomplexes seem to be the most natural choice in this setting.

Definition 1.38 Given a path-connected CW-complex X, we will refer to the quotient $\mathcal{A}(X)$ in the previous theorem as *the aspherical multicomplex* associated to X. Moreover, we denote by $\pi : |\mathcal{L}(X)| \to |\mathcal{A}(X)|$ the quotient map associated with the simplicial action of Γ_1 over $\mathcal{L}(X)$.

We are finally ready to construct our desired classifying map for a given path-connected CW-complex X:

$$\pi \circ j_X : X \longrightarrow |\mathcal{A}(X)| \simeq K(\pi_1(X), 1), \tag{1.2}$$

where $\pi : |\mathcal{L}(X)| \to |\mathcal{A}(X)|$ is the projection onto the aspherical quotient and j_X is the homotopy inverse of $S_X \circ i$ defined in Remark 1.32.

Proof of Theorem 1.3

We are now ready to show that the projection $\pi : |\mathcal{L}(X)| \to |\mathcal{A}(X)|$ induces isometric isomorphisms on bounded cohomology groups:

Theorem 1.39 [Frigerio and Moraschini, 2019, theorem 4.4.3] *Let X be a path-connected CW-complex. Then, the quotient map $\pi : |\mathcal{L}(X)| \to |\mathcal{A}(X)|$ induces isometric isomorphisms*

$$\mathrm{H}_b^n(\pi) : \mathrm{H}_b^n(|\mathcal{A}(X)|) \longrightarrow \mathrm{H}_b^n(|\mathcal{L}(X)|)$$

for every $n \geq 0$.

Hence, we can finally prove Theorem 1.3:

Proof of Theorem 1.3 Let X be a path-connected CW-complex. Then, the classifying map (1.2)

$$\pi \circ j_X : X \longrightarrow |\mathcal{A}(X)| \simeq K(\pi_1(X), 1)$$

induces for every $n \geq 0$ an isometric isomorphism on the nth bounded cohomology group (Corollary 1.31 and Theorem 1.39). □

Proof of Theorem 1.39

Recall that amenable groups are invisible to bounded cohomology (Theorem 18); hence Theorem 1.39 would be easily true if the group Γ_1 *were* amenable. However, in general Γ_1 is *not* amenable, and thus we need some more sophisticated results [Frigerio and Moraschini, 2019, theorem 4.4.1 and corollary 4.4.2] in order to prove that $\pi : |\mathcal{L}(X)| \to |\mathcal{A}(X)|$ induces isometric isomorphisms on all bounded cohomology groups. Without entering into the technical details just mentioned, it turns out that in order to prove Theorem 1.39, it is sufficient to show that Γ_1/Γ_i is amenable for every $i \geq 1$ (notice that each Γ_i is normal in Γ_1).

We spend the remaining part of this section in order to show the following:

Proposition 1.40 [Frigerio and Moraschini, 2019, corollary 4.3.11] *For every $i \geq 1$, the group Γ_1/Γ_i is solvable and thus amenable.*

Notice that we have the following normal sequence:

$$\Gamma_1/\Gamma_i \trianglerighteq \Gamma_1/\Gamma_i \trianglerighteq \cdots \trianglerighteq \Gamma_i/\Gamma_i = \{1\},$$

for every $i \geq 1$. Hence, in order to show that Γ_1/Γ_i is solvable, it is sufficient to prove that the following groups

$$\Gamma_{j-1}/\Gamma_j \cong \frac{\Gamma_{j-1}/\Gamma_i}{\Gamma_j/\Gamma_i}$$

are abelian for all $j = 1, \ldots, i - 1$. So we have to prove the following lemma:

Lemma 1.41 [Frigerio and Moraschini, 2019, corollary 4.3.10] *For every $i \geq 2$, the group Γ_{i-1}/Γ_i is abelian.*

Proof Let $i \geq 2$ and let us fix a set of representatives $\{\Delta_\alpha\}_{\alpha \in J}$ for the action of Γ_{i-1} on the set of of i-simplices of $\mathcal{L}(X)$. Moreover, let p_α be a chosen vertex of each representative Δ_α. Then, the (well-defined) map

$$\phi_\alpha : \Gamma_{i-1} \longrightarrow \pi_i(|\mathcal{L}(X)|, p_\alpha)$$
$$\gamma \longmapsto \left[\dot{S}_\alpha^i(\Delta_\alpha, \gamma(\Delta_\alpha))\right]$$

is a group homomorphisms for every $\alpha \in J$ [Frigerio and Moraschini, 2019, lemma 4.3.8].

Hence, we can consider the *direct product* of the homomorphisms ϕ_α and obtain the following homomorphism:

$$\Phi : \Gamma_{i-1} \longrightarrow \prod_{\alpha \in J} \pi_i(|\mathcal{L}(X)|, p_\alpha).$$

Notice that since $i \geq 2$, the homotopy groups $\pi_i(|\mathcal{L}(X)|, p_\alpha)$ are all abelian, and thus their direct product is also. Hence, we are reduced to show that $\ker(\Phi) = \Gamma_i$.

We begin by recalling that $\gamma \in \ker(\phi_\alpha)$ if and only if Δ_α is homotopic to $\gamma(\Delta_\alpha)$. Hence, by the minimality of $\mathcal{L}(X)$, $\gamma \in \ker(\phi_\alpha)$ if and only if $\gamma(\Delta_\alpha) = \Delta_\alpha$. This shows that the kernel of ϕ_α coincides with the stabiliser of Δ_α in Γ_{i-1}, and thus with the stabiliser of any simplex in the orbit of Δ_α (because $\ker(\phi_\alpha)$ is normal). Since the set of i-simplices of $\mathcal{L}(X)$ coincides with the union of all the orbits of the $\Delta_\alpha, \alpha \in J$, we get

$$\ker(\Phi) = \bigcap_{\alpha \in J} \ker(\phi_\alpha) = \Gamma_i;$$

hence the thesis. □

2

The Proportionality Principle via Hyperbolic Geometry

Filippo Sarti[*]

The goal of this chapter is proving the following theorem (Theorem 5):

Theorem 2.1 (Proportionality Principle (Theorem 5)) *Let $n \geq 2$ and let M be a closed and connected hyperbolic n-manifold. Then*

$$\|M\| = \frac{\mathrm{vol}(M)}{v_n},\qquad(2.1)$$

where $\|M\|$ and $\mathrm{vol}(M)$ are respectively the simplicial and the Riemannian volume of M, and v_n is a constant depending only on n.

This result, known as *Gromov's proportionality principle*, characterizes and distinguishes the hyperbolic world with respect to other Riemannian structures, and it has several remarkable consequences in the study of hyperbolic manifolds. Before starting with the proof, let us make some observations.

Remark 2.2 We point out the following:

(i) As one can immediately realise, while the left-hand side of (2.1) is a topological invariant, the right-hand side a priori strongly depends on the Riemannian structure of M. This gives us the first flavor of the rigid behaviour of hyperbolic manifolds, and it may suggest a connection with the *Mostow rigidity theorem*, which asserts that if $n \geq 3$, any two closed connected hyperbolic n-manifolds with isomorphic fundamental groups are isometric [Mostow, 1968]. In fact, a more geometrical proof of Mostow rigidity provided by Gromov [1981] is based on the notion of simplicial volume and on its proportionality relation with the Riemannian volume, namely on Theorem 2.1. We refer to Thurston [1979], Benedetti and Petronio [1992],

* Partially supported through GNSAGA-INDAM.

and Martelli [2016] for a complete account of Gromov's proof of Mostow rigidity.

(ii) The second observation concerns the constant v_n. As we will see in the proof, the first step is to prove that for any $n \geq 1$ the n-simplex of maximal volume in the n-hyperbolic space $\mathbb{H}^n$ is the *ideal regular* one. This is a mere computation for $n = 2$; it can be proved using a geometrical argument for $n = 3$ [Thurston, 1979, Chapter 7] and it is a deep result for $n \geq 4$ [Haagerup and Munkholm, 1981].

(iii) In dimension two, the only oriented closed connected surfaces admitting a hyperbolic metric are the g-tori T_g with $g \geq 2$. An easy computation shows that $v_2 = \pi$ and that $\left\| T_g \right\| = 4g - 4$ [Gromov, 1982; Frigerio, 2017]. Now, by the Gauss–Bonnet theorem, we get

$$\mathrm{vol}(T_g) = \int_{T_g} dA = - \int_{T_g} K \, dA = -2\pi \chi(T_g) = -2\pi(2-2g) = v_2 \left\| T_g \right\|,$$

where K denotes the Gaussian curvature.

Hence, in the case of oriented closed hyperbolic surfaces, Theorem 2.1 is equivalent to the Gauss-Bonnet theorem.

In the following sections we recall some tools that we will use in the proof of the main theorem. Our exposition mainly follows Martelli's book [2016].

2.1 Volume of Simplices in $\mathbb{H}^n$

As anticipated in the introduction, we need a characterization of the volume of simplices in $\mathbb{H}^n$. This is the content of the following result, which is specific to the hyperbolic space.

Theorem 2.3 *Let $n \geq 2$. There exists a constant v_n such that for any hyperbolic n-simplex Δ^n in $\overline{\mathbb{H}}^n$ we have $\mathrm{vol}(\Delta^n) \leq v_n$ and $\mathrm{vol}(\Delta^n) = v_n$ if and only if Δ^n is regular and ideal.*

The proof for $n = 2$ is a computation; for $n = 3$ we refer to Thurston [1979] and Martelli [2016, corollary 13.1.4], while for $n \geq 4$ we refer to Haagerup and Munkholm [1981].

2.2 Simplicial Volume

We recall the definition of simplicial volume (Definition 1):

Definition 2.4 Let M be an oriented connected closed topological manifold of dimension n. The *simplicial volume* of M is

$$\|M\| := \inf\left\{\sum_{i=1}^{k} |a_i|, \left[\sum_{i=1}^{k} a_i\sigma_i\right] = [M] \in H_n(M; \mathbb{R})\right\},$$

where the σ_i's are n-singular simplices in M and $[M]$ denotes the fundamental class of M.

If M is non-orientable, we set $\|M\| := \|\tilde{M}\|/2$, where $\tilde{M} \to M$ denotes the orientable double covering of M.

The simplicial volume is a topological invariant, since it depends only on the fundamental class of the manifold. One can prove even further (Proposition 3):

Proposition 2.5 [Gromov, 1982; Martelli, 2016, proposition 13.2.4]   *If* $f: M \to N$ *is a degree d map between closed and connected n-manifolds, then*

$$\|M\| \geq |d| \cdot \|N\|.$$

As a consequence, if M and N are homotopy equivalent, then $\|M\| = \|N\|$.

2.3 Straightening of Simplices

By Theorem 2.3, the ideal regular n-simplices are those of maximal volume. In order to find a correlation between Riemannian and simplicial volume, the idea is to modify, using a homotopy, the simplices of any representative of the fundamental class of M, in order to get simplices with computable volume. More precisely, one can lift any n-simplex of an n-chain to an n-simplex into the universal cover $\mathbb{H}^n$, take the convex hull of its vertices in $\mathbb{H}^n$, and then project again on M using the universal cover. Let us describe the details of this process, which is called *straightening* of simplices.

Consider an n-simplex $\sigma: \Delta^n \to \mathbb{H}^n$ with vertices $v_0, \ldots, v_n$. We define a new simplex $\sigma^{st}: \Delta^n \to \mathbb{H}^n$ as

$$\sigma^{st}(t_0, \ldots, t_n) := \frac{t_0 v_0 + \cdots + t_n v_n}{\sqrt{-\|t_0 v_0 + \cdots + t_n v_n\|^2}},$$

where the sum is that of $\mathbb{R}^{n+1}$ in the hyperboloid model and $\|\cdot\|^2$ denotes the square of the Lorentzian norm, namely

$$\|(x_0, \ldots, x_n)\|^2 = \sum_{i=0}^{n-1} x_i^2 - x_n^2.$$

Now, for a simplex $\sigma: \Delta^n \to M$, we can consider the universal cover $\pi: \mathbb{H}^n \to M$ and a lifting $\tilde{\sigma}: \Delta^n \to \mathbb{H}^n$, take its straightening $\tilde{\sigma}^{st}$ and then project again on M. Hence we can define a chain map $st_n \colon C_n(M; \mathbb{R}) \to C_n(M; \mathbb{R})$ as follows:

$$\sum_{i=1}^{k} a_i \sigma_i \longmapsto \sum_{i=1}^{k} a_i \sigma_i^{st},$$

which induces a homomorphism $st_n \colon H_n(M; \mathbb{R}) \to H_n(M; \mathbb{R})$ in homology. We have the following classical result (see, e.g., Martelli [2016, proposition 13.2.12]):

Lemma 2.6 *The map st_* is an isomorphism.*

Proof First consider the following homotopy: Given $\sigma \colon \Delta^n \to M$, we define

$$\widetilde{H} \colon \Delta^n \times [0, 1] \longrightarrow \mathbb{H}^n$$

$$(x, t) \longmapsto \frac{t\widetilde{\sigma}(x) + (1 - t)\widetilde{\sigma}^{st}(x)}{\sqrt{-\|t\widetilde{\sigma}(x) + (1 - t)\widetilde{\sigma}^{st}(x)\|^2}}.$$

Then, we can compose $\widetilde{H}$ with the universal covering projection $\pi \colon \mathbb{H}^n \to M$ and obtain the following homotopy:

$$H \colon \Delta^n \times [0, 1] \longrightarrow M$$

$$(x, t) \longmapsto \pi \circ \widetilde{H}(x, t).$$

Since $H(x, 0) = \sigma(x)$ and $H(x, 1) = \pi \circ \widetilde{\sigma}^{st}(x)$, the previous homotopy induces a chain homotopy between st and the identity. □

We now define the *abstract volume* of a straightened simplex $\sigma \colon \Delta^n \to M$ as the quantity

$$\mathrm{vol}(\sigma) := \left| \int_{\Delta^n} \omega_\sigma \right|,$$

where ω_σ is the pull-back along σ of the volume form ω of M. Hence we have the following proposition:

Proposition 2.7 *The abstract volume of a simplex $\sigma \colon \Delta^n \to M$ and the hyperbolic volume of a lifting $\widetilde{\sigma} \colon \Delta^n \to \mathbb{H}^n$ coincide.*

Proof Thanks to the following commutative diagram

we have

$$\mathrm{vol}(\widetilde{\sigma}) = \left| \int_{\widetilde{\sigma}(\Delta^n)} \pi_*(\omega) \right| = \left| \int_{\pi \circ \widetilde{\sigma}(\Delta^n)} \omega \right| = \left| \int_{\sigma(\Delta^n)} \omega \right| = \left| \int_{\Delta^n} \omega_\sigma \right| = \mathrm{vol}(\sigma),$$

where $\pi_*(\omega)$ denotes the pull-back along π of the volume form ω.   □

2.4 Efficient Cycles

The following notion is fundamental in the proof of the main Theorem 2.1. The interested reader may find a more detailed discussion in Martelli's book [2016, section 13.2.6].

Definition 2.8 Let $\varepsilon > 0$. An ε-*efficient cycle* for M is a representative $\sum_{i=1}^{k} a_i \sigma_i$ of $[M]$ such that

- each σ_i is straight;
- the sign of a_i and of $\int_{\Delta^n} \omega_{\sigma_i}$ coincide for every $i = 1, \ldots, k$;
- $\mathrm{vol}(\sigma_i) > v_n - \varepsilon$ for every $i = 1, \ldots, k$.

The existence of ε-efficient cycles for any ε is guaranteed by the following crucial proposition:

Proposition 2.9 *Let M be an oriented closed connected n-manifold with $n \geq 2$. Then for any $\varepsilon > 0$, there exists an ε-efficient cycle for M.*

Proof Let $x \in \mathbb{H}^n$ and $t > 0$. We define $\Delta_x(t)$ to be the convex hull of the image under the exponential map $\exp_x : T_x \mathbb{H}^n \to \mathbb{H}^n$ of the vertices of a regular Euclidean n-simplex with vertices at distance t from the origin. We denote by $S(t)$ the set of t-*simplices*, that is, all those simplices in $\mathbb{H}^n$ that are isometric to $\Delta_x(t)$, endowed with an ordering of their vertices. Using the definition of t-simplices, it is a standard exercise to show that $\mathrm{Isom}(\mathbb{H}^n)$ acts transitively and freely on $S(t)$ [Benedetti and Petronio, 1992, Lemma C.4.14]. If we fix $\Delta_x(t)$ as a basepoint, the orbit map for this action identifies naturally $S(t)$ with $\mathrm{Isom}(\mathbb{H}^n)$, and so $S(t)$ inherits a left $\mathrm{Isom}(\mathbb{H}^n)$-action (i.e., the natural translation of simplices in $\mathbb{H}^n$) and a right $\mathrm{Isom}(\mathbb{H}^n)$-action. By the left-invariance of the Haar measure, we get the left-invariance of the measure on $S(t)$, while the unimodularity of $\mathrm{Isom}(\mathbb{H}^n)$ implies also the right-invariance.

Consider now a closed connected hyperbolic n-manifold $M = \mathbb{H}/\Gamma$ and a base point $x_0 \in \mathbb{H}^n$. We notice that an $(n + 1)$-tuple $(g_0, \ldots, g_n) \in \Gamma^{n+1}$ defines a singular n-simplex in $\mathbb{H}^n$ with vertices $g_0(x_0), \ldots, g_n(x_0)$, while an element in Γ^{n+1}/Γ defines a singular simplex in M. Here Γ acts on Γ^{n+1} by the diagonal action induced by the left multiplication.

Consider the chain

$$c(t) = \sum_{\sigma \in \Gamma^{n+1}/\Gamma} a_\sigma(t)\sigma, \qquad (2.2)$$

where the coefficients $a_\sigma(t)$ are obtained as follows: For any $\sigma = (g_0, \ldots, g_n)$, we consider the μ_t-measure $a_\sigma^+(t)$ of the subspace $S_\sigma^+ \subset S(t)$ of positively oriented t-simplices with i^{th} vertex contained in the domain $D(g_i(x_0))$ of the Dirichlet tessellation of $\mathbb{H}^n$. Similarly, $a_\sigma^-(t)$ denotes the μ_t-measure of the set $S_\sigma^-(t)$ of negatively oriented t-simplices with the same property. Finally, we set $a_\sigma(t) = a_\sigma^+(t) - a_\sigma^-(t)$. We notice that the translation of a representative of a simplex σ by an element $g \in \Gamma$ affects neither $a_\sigma^+(t)$ nor $a_\sigma^-(t)$, and hence nor $a_\sigma(t)$, since the measure on $S(t)$ is left-invariant under the action of Isom $(\mathbb{H}^n)$.

Following Martelli [2016, lemma 13.2.17] we first prove that the sum in (2.2) is finite, namely that for any t, we have $a_\sigma(t) \neq 0$ only for a finite number of $\sigma \in \Gamma^{n+1}/\Gamma$. For any $\sigma \in \Gamma^{n+1}/\Gamma$, we consider the representative $(x_0, g_1(x_0), \ldots, g_n(x_0))$, and we denote by d and by T respectively the diameter of the Dirichlet domain $D(x_0)$ and the diameter of a t-simplex. Hence if $a_\sigma \neq 0$, then $d(x_0, g_i(x_0)) < 2d + T$ for any i. Since Γ is discrete, it follows that $a_\sigma \neq 0$ only for a finite number of σ's.

The next step is to prove that $c(t)$ is a cycle. The coefficient of any $(n-1)$-simplex $(g_0, \ldots, g_{n-1})$ appearing in $\partial c(t)$ is

$$\sum_{g \in \Gamma} -a_{(g,g_0,\ldots,g_{n-1})}(t) + a_{(g_0,g,\ldots,g_{n-1})}(t) + \ldots + (-1)^{n-1} a_{(g_0,\ldots,g_{n-1},g)}(t).$$

$$(2.3)$$

For any $i = 1, \ldots, n$, we consider the ith term in (2.3) and we prove that it vanishes. For instance, the last term is

$$\sum_{g \in \Gamma} a_{(g_0,\ldots,g_{n-1},g)}^+(t) - a_{(g_0,\ldots,g_{n-1},g)}^-(t),$$

where $a_{(g_0,\ldots,g_{n-1},g)}^+(t)$ is the μ_t-measure of the set of positively oriented t-simplices such that their first $(n-1)$-facet has the vertices lying in

$$D(g_0(x_0)), \ldots, D(g_{n-1}(x_0)),$$

while $a_{(g_0,\ldots,g_{n-1},g)}^-(t)$ is the μ_t-measure of the negatively oriented t-simplices with the same property. The two quantities coincide thanks to the right-invariance of the μ_t-measure on $S(t)$, since the two sets are obtained one from the other via the right multiplication by the (orientation-reversing) element of

Isom($\mathbb{H}^n$), which reflects $\Delta_x(t)$ along its first $(n-1)$-facet. The vanishing of the other terms can be deduced using a similar argument.

We now show that for sufficiently large t we have realized a positive multiple of the fundamental class of M. Let t be such that any t-simplex has vertices at distance bigger than $2d$. Hence, if there exists a positively oriented t-simplex with vertices in $D(g_0(x_0)), \ldots, D(g_n(x_0))$, then any straight simplex with vertices in the same cells of the tessellation is positively oriented as well. We deduce that either $a_\sigma(t) = a_\sigma^+(t)$ or $a_\sigma(t) = a_\sigma^-(t)$, which implies that

$$\int_{c(t)} \omega = \sum_{\sigma \in \Gamma^{n+1}/\Gamma} a_\sigma \int_\sigma \omega > 0$$

and therefore $[c(t)]$ must be a positive multiple of $[M]$. We denote by $\bar{c}(t)$ the rescaling of $c(t)$ such that $\bar{c}(t) = [M]$.

Finally we prove that, for any $\varepsilon > 0$, there exists a t_0 such that if $t > t_0$ then $\bar{c}(t)$ is an ε-efficient cycle for M (see also Martelli [2016, lemma 13.2.18]). We first notice that every simplex in $\bar{c}(t)$ is *d-close* to a t-simplex; namely it has vertices at distance less than d from those of a t-simplex. We suppose the existence of a sequence of simplices Δ_t which are d-close to t-simplices $\bar{\Delta}_t$ with volume smaller than $v_n - \varepsilon$. Performing an appropriate isometry, we can move the Δ_t's and the $\bar{\Delta}_t$'s so that they share the same barycentre. Hence, when $t \to \infty$, we get that both the vertices of Δ_t and the vertices of $\bar{\Delta}_t$ tend to those of an ideal regular n-simplex, while the volume is strictly less than v_n. By the continuity of the volume function, we get a contradiction. $\square$

2.5 The Proof

We will assume that M is orientable and we fix an orientation, since the proof in the non-orientable case can be mimicked by taking the double-sheeted cover $\widetilde{M} \to M$.

Proof of Theorem 2.1 We start by proving that $\mathrm{vol}(M) \le v_n \|M\|$. By Lemma 2.6, we can assume that any representative of the fundamental class of M is a sum of straightened simplices. Hence for any representative $c = \sum_{i=1}^k a_i \sigma_i$, we have

$$\mathrm{vol}(M) = \int_c \omega = \sum_{i=1}^k a_i \int_{\Delta^n} \omega_{\sigma_i} \le \sum_{i=1}^k |a_i| \left| \int_{\Delta^n} \omega_{\sigma_i} \right|$$
$$= \sum_{i=1}^k |a_i| \, \mathrm{vol}(\sigma_i)$$

$$= \sum_{i=1}^{k} |a_i| \, \mathrm{vol}(\widetilde{\sigma}_i)$$

$$\leq \sum_{i=1}^{k} |a_i| v_n.$$

Hence, taking the infimum over all representatives, we get that

$$\mathrm{vol}(M) \leq \inf \left\{ \sum_{i=1}^{k} |a_i| v_n, \left[\sum_{i=1}^{k} a_i \sigma_i \right] = [M] \right\} = v_n \|M\|.$$

Let us move to the converse inequality, namely $\mathrm{vol}(M) \geq v_n \|M\|$. Thanks to Proposition 2.9, for any $\varepsilon > 0$, we can consider an ε-efficient cycle $c = \sum_{i=1}^{k} a_i \sigma_i$. Hence we have

$$\mathrm{vol}(M) = \int_c \omega = \sum_{i=1}^{k} a_i \int_{\Delta^n} \omega_{\sigma_i} = \sum_{i=1}^{k} |a_i| \left| \int_{\Delta^n} \omega_{\sigma_i} \right| > \sum_{i=1}^{k} |a_i| (v_n - \varepsilon)$$

and hence, passing to the infimum, we have that $\mathrm{vol}(M) > (v_n - \varepsilon) \|M\|$. Finally, since the last inequality holds for any $\varepsilon > 0$, we get the desired inequality. $\qquad\square$

3

Positivity of Simplicial Volume via Barycentric Techniques

Shi Wang

In this chapter, we extend the positivity result from Chapter 2 (Theorem 2.1) to the context of certain non-positively-curved manifolds. While the geodesic straightening still works for strictly negatively curved manifolds, estimating the volume of the straightened simplices is generally very difficult. However, for a large class of non-positively-curved manifolds, the *barycentric straightening* turns out to be very useful. Indeed, the Jacobian of the barycentric straightened simplices can be estimated and shown to be uniformly bounded. The positivity of the simplicial volume then follows immediately.

Throughout the chapter, we use the following notation: Let X be a Hadamard space, that is, a simply connected, non-positively-curved manifold, $\Gamma < \mathrm{Isom}(X)$ be a torsion-free cocompact lattice, and $M = \Gamma \backslash X$ be the quotient manifold. Denote $\dim M = n$.

3.1 Results and Examples

We have already seen in Chapter 2 (Theorem 2.1) that the simplicial volume of a hyperbolic manifold is proportional to its hyperbolic volume, with the explicit multiplicative constant depending only on the dimension n. In particular, the simplicial volume is positive. It is then natural to ask, for what type of non-positively-curved manifolds is the simplicial volume positive? Before presenting the results, we first give the definition of a (geometric) rank, which yields the classification result for all closed non-positively-curved manifolds.

Definition 3.1 Let M be a non-positively-curved manifold. For any non-zero vector $v \in TM$, we define the *rank*$^{(+)}$ of v to be the dimension of the space of all parallel Jacobi fields along the geodesic ray formed by v. We say M is *rank one* if there exists a rank one vector on TM, and it is *higher rank* otherwise.

In the above definition, we recall that a vector field $J(t)$ on a geodesic (ray) γ is called a *Jacobi field* if it satisfies the Jacobi equation $J'' + R(J, \gamma')\gamma' = 0$. Geometrically, it gives a geodesic variation in the direction of the vector field.

The following rank rigidity theorem classifies all closed higher rank manifolds.

Theorem 3.2 [Ballmann, 1985; Burns and Spatzier, 1987] *If M is a closed non-positively-curved manifold of higher rank, then $\tilde{M}$ is either a Riemannian product of non-positively-curved manifolds, or an irreducible higher rank symmetric space of non-compact type.*

Using the classification result, we provide in the following diagram some known results regarding the positivity of the simplicial volume:

$$K \leq 0 \begin{cases} \text{Rank one} \begin{cases} \text{Negatively curved: } \|M\| > 0 \\ \text{Presence of zero sectional curvature: Mysterious} \end{cases} \\ \text{Higher rank} \begin{cases} \text{Locally a product: Understood by factors} \\ \text{Irreducible locally symmetric: } \|M\| > 0 \end{cases} \end{cases}$$

The negatively curved case is due to Inoue and Yano [1982] and the irreducible locally symmetric one is due to Lafont and Schmidt [2006] and Bucher [2007].

Thus the only mysterious cases are those rank one manifolds that contain zero sectional curvatures somewhere. As a motivation, we consider the following two examples of (generalized) graph manifolds, since they illustrate quite nicely the general phenomenon on how the curvature (geometric quantity) interacts with the simplicial volume (topological quantity).

Figure 3.1 consists of two hyperbolic surfaces Σ_1 and Σ_2 with one puncture each, such that the cusps have been truncated and smoothly and symmetrically tapered to a flat metric in a neighbourhood of their round circle boundaries $\partial \Sigma_i$. We form M by gluing $\Sigma_1 \times S^1$ to $S^1 \times \Sigma_2$ by the identity isometry along the flat boundary torus $T^2 = \partial \Sigma_1 \times S^1 \cong S^1 \times \partial \Sigma_2$, but switching the surface and circle factors. The resulting 3-manifold is non-positively-curved and has zero simplicial volume. One geometric point of view of this is that it has "too many" zero curvature planes.

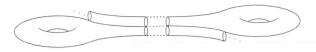

Figure 3.1 $\|M\| = 0$

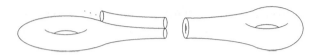

Figure 3.2 $\|M\| > 0$

Figure 3.2 consists of one identical copy of Σ_1, as in Figure 3.1, together with a hyperbolic 3-manifold N with one torus cusp truncated and tapered in a similar way to have a totally geodesic boundary of flat T^2. We form M by gluing isometrically $\Sigma_1 \times S^1$ to N along the flat boundary torus. The resulting 3-manifold also has non-positive curvature, but the simplicial volume is positive. The geometric intuition here is that it has enough negative sectional curvature (on the N part).

Based on the two examples illustrated above, it is shown in general that:

Theorem 3.3 [Connell and Wang, 2020]   *If there exists a point $x \in M$ so that every vector in $T_x^1 M$ is rank$^{(+)}$ one, then the simplicial volume $\|M\| > 0$.*

In particular, this one-point rank condition includes all non-positively-curved manifolds with a point of negative curvature, and this actually explains why the simplicial volume in Figure 3.2 is positive.

Corollary 3.4 *If there exists a point $x \in M$ so that every 2-plane in $T_x M$ is negatively curved, then the simplicial volume $\|M\| > 0$.*

Along the same line, we particularly mention a well-known open question attributed to Gromov [Savage, 1982] (see also [Gromov, 1982, p. 11]):

Question 3.5 (Gromov) Is it true that any closed manifold of non-positive sectional curvature and negative Ricci curvature has $\|M\| > 0$?

In dimension 3, Theorem 3.3 together with a Bochner type inequality gives a positive answer to the question. In higher dimensions, the question is also partially answered by the following theorem:

Theorem 3.6 [Connell and Wang, 2020]   *If there exists $x \in M$ such that any vector $v_x \in T_x^1 M$ satisfies $\mathrm{Ric}_{\lfloor \frac{n}{4} \rfloor + 1}(v_x, v_x) < 0$, then the simplicial volume $\|M\| > 0$.*

The notion of k-Ricci curvature is defined as follows:

Definition 3.7 For M non-positively-curved and $u, v \in T_x M$, the k-*Ricci curvature* is given by

$$\mathrm{Ric}_k(u, v) = \sup_{\substack{V \subset T_x M \\ \dim V = k}} \mathrm{Tr}\, R(u, \cdot, v, \cdot)|_V.$$

3.2 Straightening and Local Straightening

One main technique to prove a positivity result of the simplicial volume is the straightening method. First, we recall the following general definition of a straightening, which can be treated as a generalization of the geodesic straightening:

Definition 3.8 [Lafont and Schmidt, 2006] Let Γ be a torsion-free cocompact lattice in $\mathrm{Isom}(X)$, and let $C_\bullet(X)$ be the real singular chain complex of X. Equivalently, $C_k(X)$ is the free $\mathbb{R}$-module generated by $C^0(\Delta^k, X)$, the set of singular k-simplices in X, where Δ^k is equipped with some fixed Riemannian metric. We say a collection of maps $st_k \colon C^0(\Delta^k, X) \to C^0(\Delta^k, X)$ is a *straightening* if it satisfies the following conditions:

1. the maps st_k are Γ-equivariant,
2. the maps $st_\bullet$ induce a chain map $st_\bullet \colon C_\bullet(X; \mathbb{R}) \to C_\bullet(X; \mathbb{R})$ that is Γ-equivariantly chain homotopic to the identity,
3. the image of st_n lies in $C^1(\Delta^n, X)$, that is, the top dimensional straightened simplices are C^1,
4. there exists a constant C depending on M and the chosen Riemannian metric on Δ^n, such that for any $f \in C^0(\Delta^n, X)$, and corresponding straightened simplex $st_n(f) \colon \Delta^n \to X$, there is a uniform upper bound on the Jacobian of $st_n(f)$:

$$|\mathrm{Jac}(st_n(f))(\delta)| \le C$$

for all $\delta \in \Delta^n$.

Remark 3.9 A usual definition of straightening would just require (1)–(3), and the condition (4) assures straightened simplices have uniformly bounded volume, which is essential to getting the positivity of the simplicial volume. See below for the theorem.

By using the same idea as in the proof of Theorem 2.3, we have the following:

Theorem 3.10 [Thurston, 1979; Lafont and Schmidt, 2006] *If M admits a straightening, then the simplicial volume of M is positive.*

In fact, we can relax the definition of a straightening to become "local" while still obtaining the positivity result.

Definition 3.11 [Connell and Wang, 2020]   We say a collection of maps $st_k \colon C^0(\Delta^k, X) \to C^0(\Delta^k, X)$ is a *straightening subordinated to U* for some $U \subset M$, if it satisfies the following conditions:

1-3. same as in Definition 3.8,
  4'. there exists a constant C depending on X, U, and the chosen Riemannian metric on Δ^n, such that for any pair $(f, \delta) \in C^0(\Delta^n, X) \times \Delta^n$ satisfying $st_n(f)(\delta) \in p^{-1}(U)$, we have a uniform upper bound on the Jacobian of $st_n(f)$ at δ:

$$|\operatorname{Jac}(st_n(f))(\delta)| \leq C,$$

  where $st_n(f) \colon \Delta^n \to X$ is the corresponding straightened simplex of f and $p \colon X \to M$ is the covering map.

The local version of Theorem 3.10 now becomes

Theorem 3.12 [Connell and Wang, 2020]   *If M admits a straightening subordinated to some non-empty open set U, then the simplicial volume of M is positive.*

Proof   We choose a non-trivial smooth bump function $\phi(x)$ on M, such that $0 \leq \phi \leq 1$, and $\phi(x) = 0$ for all $x \notin U$. Let $\sum_{i=1}^{\ell} a_i \sigma_i$ be a singular chain representing the fundamental class $[M]$ in real coefficients, and $st(\sigma_i)$ be the straightened simplex of σ_i on M, with lift $\widetilde{st(\sigma_i)}$ on the universal cover X. We have

$$\int_M \phi(x)dV = \int_{[\sum a_i \sigma_i]} \phi(x)dV = \int_{[\sum a_i st(\sigma_i)]} \phi(x)dV \tag{3.1}$$

$$\leq \sum_{i=1}^{\ell} |a_i| \cdot \left| \int_{\widetilde{st(\sigma_i)}} \tilde{\phi}(x)d\tilde{V} \right| \tag{3.2}$$

$$= \sum_{i=1}^{\ell} |a_i| \cdot \left| \int_{\widetilde{st(\sigma_i)} \cap p^{-1}(U)} \tilde{\phi}(x)d\tilde{V} \right| \tag{3.3}$$

$$\leq \sum_{i=1}^{\ell} |a_i| \cdot \left| \int_{(\widetilde{st(\sigma_i)})^{-1}(U)} \phi(st(\sigma_i)(\delta))| \operatorname{Jac}(st(\sigma_i))(\delta)|dV_\Delta \right| \tag{3.4}$$

$$\leq \sum_{i=1}^{\ell} |a_i| \cdot C \operatorname{vol}(\Delta^n), \tag{3.5}$$

where equation (3.1) follows from (2) of Definition 3.11, inequality (3.2) lifts to the universal cover X, equation (3.3) uses the support of ϕ, inequality (3.4) pulls the integral back on Δ^n, and inequality (3.5) follows from Items (3) and (4) of Definition 3.11.

By taking the infimum over all fundamental class representatives $\sum_{i=1}^{\ell} a_i \sigma_i$, we have

$$\|M\| \geq \frac{\int_M \phi(x) dV}{C \, \mathrm{vol}(\Delta^n)} > 0. \qquad \square$$

3.3 Barycentric Straightening

The barycentric straightening was introduced by Lafont and Schmidt [2006] (based on the barycentre method originally developed by Besson, Courtois, and Gallot [Besson et al., 1995]) to show the positivity of simplicial volume for most locally symmetric spaces of non-compact type.

Briefly speaking, for any k-simplex on X, the corresponding $(k + 1)$ vertices form their Patterson–Sullivan measures. These measures can be viewed as $(k + 1)$ vertices in the affine space $\mathcal{M}(\partial_\infty X)$ of all measures supported on the boundary of X. By the affine structure, we can fill up a simplex on the space of measures by taking the linear combinations of the measures. Finally, applying the barycentre map, it gives back a simplex on X. Such a simplex is then defined to be the barycentrically straightened simplex. Now we describe them in detail.

3.3.1 Patterson–Sullivan Measures

As a first step, we want to assign to any point $x \in X$ a measure supported on the boundary at infinity $\partial_\infty X$. In fact, there is very much flexibility to choose such measures, and in most cases, the choice will not affect later estimates (except only in higher rank symmetric spaces), so that it makes no difference. For example, one can choose the Lebesgue measures, or harmonic measures at x, as long as the assignment is Γ-equivariant. However, in the case of higher rank symmetric spaces, we need to further require that the resulting measures support on a "regular" set (a special realisation of the Furstenberg boundary) inside the entire visual boundary, in order for the later estimates to work. For this reason, we particularly introduce the Patterson–Sullivan measures here.

Definition 3.13 We call a family of finite Borel measures $\{\mu_x\}_{x \in X}$ supported on $\partial_\infty X$ the *Patterson–Sullivan measures*, if it satisfies:

1. μ_x is Γ-equivariant, that is, $\gamma_* \mu_x = \mu_{\gamma x}$ for all $x \in X$ and $\gamma \in \Gamma$, and

2. $\frac{d\mu_x}{d\mu_y}(\theta) = e^{hB(x,y,\theta)}$, for all x, $y \in X$, and $\theta \in \partial_\infty X$,

where h is the volume growth entropy of M, and $B(x, y, \theta)$ is the Busemann function on X defined to be $B(x, y, \theta) = \lim_{t \to \infty}(d_X(y, \gamma_\theta(t)) - t)$, such that γ_θ is the geodesic ray from x to θ.

In the case of higher rank symmetric spaces, Albuquerque [1999] showed the existence and uniqueness of the Patterson–Sullivan measures, and the support of the measures lies in the most regular directions called the Furstenberg boundary. In the case of geometric rank ones, Knieper [1997] proved the existence and uniqueness.

3.3.2 Barycentres

As a second step, we illustrate how we obtain from a measure supported at $\partial_\infty X$ a barycentre in X. Let ν be any finite Borel measure fully supported on $\partial_\infty X$ in case of geometric rank one; in case of higher rank, ν is fully supported on $\partial_F X$, where $\partial_F X$ is identified with the G-orbits of a boundary point that corresponds to the tangent vector dual to the sum of positive restricted roots in $\mathfrak{g}$ (here $G = \mathrm{Isom}^0(X)$ and $\mathfrak{g}$ is its Lie algebra). If we set $B(x, \theta) := B(p, x, \theta)$ for some fixed basepoint $p \in X$, and by taking the integral of $B(x, \theta)$ with respect to ν, we obtain a function

$$x \longmapsto \mathcal{B}_\nu(x) := \int_{\partial_\infty X} B(x, \theta) d\nu(\theta).$$

One can show that if ν satisfies the above-mentioned support condition, then $\mathcal{B}_\nu$ is strictly convex, and if ν is further a linear sum of the Patterson–Sullivan measures, then $\mathcal{B}_\nu$ attains a unique minimum in X, which we denote by $\mathrm{bar}(\nu)$. It is not difficult to see that $\mathrm{bar}(\nu)$ does not depend on the choice of basepoint p.

3.3.3 The Explicit Construction

Now we are ready to define the barycentric straightening map. We denote by Δ_s^k the standard spherical k-simplex in the Euclidean space, that is

$$\Delta_s^k = \left\{ (a_1, \ldots, a_{k+1}) \mid a_i \geq 0, \sum_{i=1}^{k+1} a_i^2 = 1 \right\} \subseteq \mathbb{R}^{k+1},$$

with the induced Riemannian metric from $\mathbb{R}^{k+1}$, and with ordered vertices $(e_1, \ldots, e_{k+1})$. Given any singular k-simplex $f: \Delta_s^k \to X$, with ordered vertices $V = (x_1, \ldots, x_{k+1}) = (f(e_1), \ldots, f(e_{k+1}))$, we define the k-straightened simplex

$$st_k(f): \Delta_s^k \longrightarrow X$$

$$st_k(f)(a_1, \ldots, a_{k+1}) := \mathrm{bar}\left(\sum_{i=1}^{k+1} a_i^2 v_{x_i}\right),$$

where $v_{x_i} = \mu_{x_i}/\|\mu_{x_i}\|$ is the normalized Patterson–Sullivan measure at x_i. We notice that $st_k(f)$ is determined by the (ordered) vertex set V, and we denote $st_k(f)(\delta)$ by $st_V(\delta)$, for $\delta \in \Delta_s^k$.

Proposition 3.14 *The above-defined barycentric straightening satisfies* (1)–(3) *of Definition 3.8 (hence also Definition 3.11). Moreover, the Jacobian of the barycentric straightening map can be estimated as*

$$|\mathrm{Jac}(st_n(f))(\delta)| \le 2^n \cdot \frac{\det(H_{\delta,V})^{1/2}}{\det(K_{\delta,V})}, \tag{3.6}$$

for all $f \in C^0(\Delta_s^n, X)$, $\delta = (a_1, \ldots, a_{n+1}) \in \Delta_s^n$, where

$$H_{\delta,V} = \int_{\partial_\infty X} \left(dB_{(st_V(\delta),\theta)}\right)^2 d\left(\sum_{i=1}^{n+1} a_i^2 v_{x_i}\right)(\theta),$$

$$K_{\delta,V} = \int_{\partial_\infty X} DdB_{(st_V(\delta),\theta)} d\left(\sum_{i=1}^{n+1} a_i^2 v_{x_i}\right)(\theta).$$

Proof The proof is verbatim as in the case of higher rank symmetric spaces [Lafont and Schmidt, 2006]. $\qquad\square$

Remark 3.15 The two symmetric bilinear forms are essentially the average of $dB \otimes dB$ and $\mathrm{Hess}(B)$ over some probability measure which is a finite sum of the Patterson–Sullivan measures. Using the non-positive curvature, the Busemann functions are convex, and one obtains that the bilinear forms are positive definite. The quotient of the determinants that appears on the right side of inequality (3.6) has been estimated by Besson–Courtois–Gallot [Besson et al., 1995] in rank one symmetric spaces (sharp) and by Connell–Farb [Connell and Farb, 2003] in higher rank (non-sharp).

3.4 Jacobian Estimates

Motivated by inequality (3.6), if we can bound the quotient $\frac{\det(H_{\delta,V})^{1/2}}{\det(K_{\delta,V})}$ uniformly (or at least locally on some $U \subseteq M$), then the barycentric straightening is indeed a straightening (or straightening subordinated to U), which then implies the positivity of the simplicial volume according to Theorems 3.10

and 3.12. More precisely, we can summarize the above discussion into the following theorem, after rewriting the bilinear forms in the following way:

$$H_{x,\nu} = \int_{\partial_\infty X} \left(dB_{(x,\theta)} \right)^2 d\nu(\theta),$$

$$K_{x,\nu} = \int_{\partial_\infty X} Dd B_{(x,\theta)} d\nu(\theta).$$

Theorem 3.16 *Given M, a closed non-positively-curved manifold, if there exists a constant C and a non-empty open set $U \subset M$ such that, for any probability measure ν which is a finite linear sum of the Patterson–Sullivan measures, and any $x \in X$ whose natural projection $p(x) \in U$, we have*

$$\frac{\det(H_{x,\nu})^{1/2}}{\det(K_{x,\nu})} \le C, \tag{3.7}$$

then $\|M\| > 0$.

In the rest of the section, we attempt to illustrate (at least philosophically) how to obtain the uniform bound (3.7) for certain cases of non-positively-curved manifolds whose negative curvature "dominates" in some sense.

3.4.1 Locally Symmetric Manifolds

Now we restrict to the case of higher rank locally symmetric space; our approach follows from Connell and Farb [2003] and Lafont and Wang [2019]. We assume for simplicity $X = G/K$ is irreducible, where G is a connected, simple Lie group with finite center, and K is its maximal compact subgroup. One typical such example is given by $X = \mathrm{SL}_m(\mathbb{R})/\mathrm{SO}(m)$.

Our goal is to analyze the eigenvalues of $H_{x,\nu}$ and $K_{x,\nu}$ (for simplicity, we just write H and K). Indeed, since the symmetric tensors $dB \otimes dB$ and DdB have eigenvalues bounded from above, after taking the average, the eigenvalues remain bounded. Therefore, in order to obtain a uniform bound as in (3.7), it suffices to show the following eigenvalue matching property:

Eigenvalue Matching Property: For any small eigenvalue λ_i of K, there is a distinct pair of eigenvalues μ_i, μ_i' of H, which cancels with λ_i, that is, $\mu_i, \mu_i' = O(\lambda_i)$. (The threshold value for an eigenvalue to be small can be chosen arbitrarily to depend on the type of symmetric space but is independent on the choice of x and ν that defines H and K.)

First, we investigate how many small eigenvalues K can have at most. To do this, we need to introduce the following notion of k-trace:

Definition 3.17 Let L be a positive definite symmetric bilinear form on V^n, and $k \leq n$ be any natural number. We define the k-*trace* of L to be

$$\mathrm{Tr}_k(L) := \inf_{\substack{V_0 \subseteq V \\ \dim V_0 = k}} \mathrm{Tr}(L|_{V_0}).$$

Equivalently, it is the sum of the first k-eigenvalues of L.

One nice property of this notion is that it is superadditive, that is,

$$\mathrm{Tr}_k(L_1 + L_2) \geq \mathrm{Tr}_k(L_1) + \mathrm{Tr}_k(L_2).$$

So, in our case, we obtain that

$$\mathrm{Tr}_k(K) \geq \int_{\partial_\infty X} \mathrm{Tr}_k(DdB_{(x,\theta)}) d\nu(\theta).$$

Now we use the geometry of the symmetric space, and since ν supports on the Furstenberg boundary, we only need to consider $\theta \in \partial_F X$; so $DdB_{(x,\theta)}$ has exactly r zero eigenvalues, where r is the rank of the symmetric space X, and so $\mathrm{Tr}_{r+1}(DdB_{(x,\theta)}) \geq \varepsilon_0$ holds for some constant ε_0 depending on X. This, together with the superadditivity, immediately implies the following:

Lemma 3.18 *K has at most r small eigenvalues, where r is the rank of the symmetric space X.*

Next, we consider the matching for only a single small eigenvalue of K. If we denote by $v \in T_x^1 X$ any vector which corresponds to this eigenvalue, the following lemma gives a description of the feasible set of vectors which can produce comparably small (eigen)values of H.

Lemma 3.19 *Let $v \in T_x^1 X$ be such that it is most singular (in the sense of maximal isotropy subgroup) in its ρ-neighbourhood. Then there exists a constant $C(\rho)$ such that, for any $u \in \mathrm{Null}\left(R(v, \cdot, v, \cdot)\right)^\perp := F_v$, we have*

$$H(u, u) \leq C(\rho) K(v, v).$$

Proof The original proof goes back to Connell and Farb [2003] using Lie algebra computations, but a general approach has been obtained in Connell and Wang [2019, lemma 5.3], with a slight modification required here in adaptation to symmetric spaces. $\square$

Finally, to arrange a matching for all small eigenvalues, we observe that the eigenvectors of these small eigenvalues (assume the number is k, $k \leq r$) will be almost inside a flat. By perturbation, we may assume these directions $v_1, \ldots, v_k$ span an almost orthonormal frame entirely inside the flat $\mathcal{F}$, and they are most singular within a fixed small neighborhood. Now by Lemma 3.19, if we can find distinct pairs $u_i, u'_i \in F_{v_i}$ for each v_i, such that $\{u_1, \ldots, u_k, u'_1, \ldots, u'_k\}$ spans an orthonormal frame (necessarily in $\mathcal{F}^\perp$ by definition of F_v), then the eigenvalue matching property holds. It turns out that this boils down to a certain combinatorial problem that can be solved by the generalized Hall's marriage theorem (see Lafont and Wang [2019]). In other words, we have the following sufficient condition for the eigenvalue matching property:

Proposition 3.20 [Lafont and Wang, 2019] *Let $\{v_1, \ldots, v_r\}$ be an almost orthonormal r-frame in $\mathcal{F}$. If for any subcollection $\{v_{i_1}, \ldots, v_{i_k}\} \subseteq \{v_1, \ldots, v_r\}$, the following dimension inequality holds*

$$\dim(F_{v_{i_1}} \oplus F_{v_{i_2}} \oplus \cdots \oplus F_{v_{i_k}}) \geq 2k,$$

then the eigenvalue matching property holds.

We now illustrate the eigenvalue matching in the following two examples where $G = \mathrm{SL}_3(\mathbb{R})$ and $\mathrm{SL}_4(\mathbb{R})$.

Example 3.21 $X = \mathrm{SL}_3(\mathbb{R})/\mathrm{SO}(3)$: The matching fails. The problem is when $\{v_1, v_2\}$ takes the entire orthonormal frame of the two-dimensional flat $\mathcal{F}$. By definition, F_{v_i} is either $\mathcal{F}^\perp$ if v_i is regular, or a two-dimensional subspace in $\mathcal{F}^\perp$ if v_i is singular. We have that

$$F_{v_1} \oplus F_{v_2} \subseteq \mathcal{F}^\perp,$$

whose dimension is at most 3. Thus the inequality in Proposition 3.20 fails.

Example 3.22 $X = \mathrm{SL}_4(\mathbb{R})/\mathrm{SO}(4)$: The matching works. We consider the worst case where we have three small eigenvalues, whose eigenvectors correspond to the most singular directions and therefore the feasible sets are the smallest possible. In fact this cannot happen, but even so we can show there exists a matching, let alone other cases. For any subcollection, without loss of generality, it is $\{v_1\}$, $\{v_1, v_2\}$, or $\{v_1, v_2, v_3\}$. We can check that

1. $\{v_1\}$: $\dim F_{v_1} = \dim(\mathrm{SL}_3(\mathbb{R})/\mathrm{SO}(3) \times \mathbb{R})^\perp = 3 \geq 2 \cdot 1$,
   or that $\dim F_{v_1} = \dim(\mathbb{H}^2 \times \mathbb{H}^2 \times \mathbb{R})^\perp = 4 \geq 2 \cdot 1$,
2. $\{v_1, v_2\}$: $\dim(F_{v_1} \oplus F_{v_2}) = \dim(\mathbb{H}^2 \times \mathbb{R}^2)^\perp = 5 \geq 2 \cdot 2$,
3. $\{v_1, v_2, v_3\}$: $\dim(F_{v_1} \oplus F_{v_2} \oplus F_{v_3}) = \dim(\mathcal{F}^\perp) = 6 \geq 2 \cdot 3$.

Therefore, the dimension inequality in Proposition 3.20 holds. Moreover, we can see exactly how the pairings work by writing them explicitly.

We identify the tangent space $T_x X \cong \mathfrak{p}$, where $\mathfrak{p}$ is the symmetric part under the Cartan decomposition, which consists of traceless 4×4 symmetric matrices, and the flat $\mathcal{F}$ corresponds to the diagonal ones. Suppose the three most singular directions v_1, v_2, v_3 are (up to a scale) given by

$$v_1 = \begin{bmatrix} -3 & & & \\ & 1 & & \\ & & 1 & \\ & & & 1 \end{bmatrix}, v_2 = \begin{bmatrix} 1 & & & \\ & -3 & & \\ & & 1 & \\ & & & 1 \end{bmatrix}, v_3 = \begin{bmatrix} 1 & & & \\ & 1 & & \\ & & -3 & \\ & & & 1 \end{bmatrix}.$$

Then their corresponding feasible sets F_{v_i} are as follows:

$$F_{v_1} = \mathrm{span}\left\{ \begin{bmatrix} 0 & 1 & & \\ 1 & 0 & & \\ & & 0 & \\ & & & 0 \end{bmatrix}^{(*)}, \begin{bmatrix} 0 & & 1 & \\ & 0 & & \\ 1 & & 0 & \\ & & & 0 \end{bmatrix}, \begin{bmatrix} 0 & & & 1 \\ & 0 & & \\ & & 0 & \\ 1 & & & 0 \end{bmatrix}^{(*)} \right\},$$

$$F_{v_2} = \mathrm{span}\left\{ \begin{bmatrix} 0 & 1 & & \\ 1 & 0 & & \\ & & 0 & \\ & & & 0 \end{bmatrix}, \begin{bmatrix} 0 & & & \\ & 0 & 1 & \\ & 1 & 0 & \\ & & & 0 \end{bmatrix}^{(*)}, \begin{bmatrix} 0 & & & \\ & 0 & & 1 \\ & & 0 & \\ & 1 & & 0 \end{bmatrix}^{(*)} \right\},$$

$$F_{v_3} = \mathrm{span}\left\{ \begin{bmatrix} 0 & & 1 & \\ & 0 & & \\ 1 & & 0 & \\ & & & 0 \end{bmatrix}^{(*)}, \begin{bmatrix} 0 & & & \\ & 0 & 1 & \\ & 1 & 0 & \\ & & & 0 \end{bmatrix}, \begin{bmatrix} 0 & & & \\ & 0 & & \\ & & 0 & 1 \\ & & 1 & 0 \end{bmatrix}^{(*)} \right\}.$$

Now by carefully choosing each u_i, u_i' as indicated by $(*)$ above, we obtain a matching.

3.4.2 Geometric Rank One Manifolds

Now we turn to the case of geometric rank one. Note that not every rank one manifold has positive simplicial volume (see Figure 3.1). Thus, we propose two sufficient conditions that are both local, one with a rank condition and the other with a curvature condition. We also give a sketch of the proofs of Theorem 3.3 and 3.6.

Under a single point rank$^+$ one condition. This case is relatively easier, and essentially due to the original estimate of Besson–Courtois–Gallot [Besson et al., 1995].

Proof of Theorem 3.3 We set u as a non-negative function on M that detects rank$^+$ one vectors, namely, let $u(x) := \inf_{v_\theta \in T^1_{\bar{x}} M} \mathrm{Tr}_2\, Dd B_{(\bar{x}, \theta)}$, where $\bar{x} \in X$ is a lift of x. Note that $u(x) > 0$ if and only if all vectors on x are rank$^+$ one. Thus, by assumption, there exists $p \in M$ such that $u(p) > 0$. We choose a neighborhood U of p such that $u(x) \geq \varepsilon_0$ for all $x \in U$.

According to Theorem 3.16, we just need to obtain a uniform bound as in (3.7). Note that under a proper choice of diagonalizing orthonormal basis $e_1, e_2, \ldots, e_n$ so that e_1 is the unit vector at $\bar{x}$ pointing toward θ, we can write the bilinear forms in the following matrix forms:

$$(d B_{(\bar{x}, \theta)})^2 = \begin{bmatrix} 1 & 0 \\ 0 & 0^{(n-1)} \end{bmatrix}, \text{ and}$$

$$Dd B_{(\bar{x}, \theta)} \geq u(x) \begin{bmatrix} 0 & 0 \\ 0 & I^{(n-1)} \end{bmatrix}.$$

Therefore, we have for any $(\bar{x}, \theta)$ with $p(\bar{x}) = x \in U$ that

$$(d B_{(\bar{x}, \theta)})^2 + \frac{1}{u(x)} Dd B_{(\bar{x}, \theta)} \geq I^{(n)}.$$

Hence, after integrating with respect to any probability measure ν, the following holds:

$$H_{\bar{x}, \nu} + \frac{1}{u(x)} K_{\bar{x}, \nu} \geq I^{(n)}.$$

This inequality is independent on ν. We now apply the following lemma from Besson–Courtois–Gallot [Besson et al., 1995, proposition B.1] on $H_{\bar{x}, \nu}$ and $\frac{1}{u(x)} K_{\bar{x}, \nu}$.

Lemma 3.23 *Let H and K be two $n \times n$ $(n \geq 3)$ matrices, where K is positive definite and H is positive semidefinite. If $H + K \geq I$ and $tr(H) = 1$, then*

$$\frac{\det(H)^{1/2}}{\det(K)} \leq \frac{n^{n/2}}{(n-1)^n}.$$

Thus, we obtain

$$\frac{\det(H_{\bar{x}, \nu})^{1/2}}{\det(K_{\bar{x}, \nu})} \leq \frac{1}{u^n(x)} \cdot \frac{n^{n/2}}{(n-1)^n},$$

and the desired uniform bound on U then follows:

$$\frac{\det(H_{\bar{x},v})^{1/2}}{\det(K_{\bar{x},v})} \leq \frac{n^{n/2}}{\varepsilon_0^n (n-1)^n}.$$

This completes the proof in view of Theorem 3.16. $\qquad\square$

Under a single point negative k-Ricci condition. To prove Theorem 3.6 in this case, we need to obtain a similar eigenvalue matching as in the case of symmetric spaces. We are going to establish parallel results to Lemmas 3.18 and 3.19. Following the same notations as above, $K := K_{x,v}$ and $H := H_{x,v}$ for some $x \in U$ and v is the finite sum of Patterson–Sullivan measures. Again, we first investigate how many small eigenvalues of K one can have at most.

Lemma 3.24 *If M satisfies $\mathrm{Ric}_{k+1} < 0$, then there exists an $\varepsilon > 0$ that only depends on $\widetilde{M} := X$, such that the number of eigenvalues of K which is $< \varepsilon$ is no more than k.*

Proof Use the same idea as in Lemma 3.18. See Connell and Wang [2019, corollary 1]. $\qquad\square$

In particular, under the assumption of Theorem 3.6, there are at most $\lfloor \frac{n}{4} \rfloor$ small eigenvalues. Next, we consider the matching for a single eigenvalue.

Lemma 3.25 *Suppose M has uniformly negative $(\lfloor \frac{n}{4} \rfloor + 1)$-Ricci curvature in a neighbourhood U of x. Then there is a constant C that depends on (M, g) and U, such that for all $v \in T_x^1 M$, there is a subspace $F_v \subseteq T_x M$ of dimension at least $\lfloor \frac{3n}{4} \rfloor$, and the inequality*

$$dB_{(x,\theta)}^2(u, u) \leq C \left(DdB_{(x,\theta)}(v, v) \right)^{2/3}$$

holds for all $u \in F_v$.

By integrating the above inequality with respect to v, and applying the Hölder inequality, it follows that

Eigenvalue Matching Property: For any small eigenvalue λ of K, there is a collection of eigenvalues $\{\mu_1, \ldots, \mu_{\lfloor \frac{3n}{4} \rfloor}\}$ (counting multiplicity) of H, which satisfies $\mu_i - O(\lambda^{2/3})$ for all $1 \leq i \leq \lfloor \frac{3n}{4} \rfloor$.

Finally, we are ready to prove Theorem 3.6.

Proof of Theorem 3.6 For any $x \in U$, and any finite sum of Patterson–Sullivan measures ν, we give a uniform upper bound as in (3.7). We denote $\lambda_1 \leq \lambda_2 \leq \cdots \leq \lambda_n$ the eigenvalues of K and $\mu_1 \leq \mu_2 \leq \cdots \leq \mu_n \leq 1$ the eigenvalues of H. By Lemma 3.24, we have $\lambda_i \geq \varepsilon$, which for some ε depends on U and X, and for all $i > k$ where $k = \lfloor \frac{n}{4} \rfloor$. By the Eigenvalue matching property (essentially due to Lemma 3.25), we have $\mu_i = O(\lambda_1^{2/3})$ for all $1 \leq i \leq 3k$. Thus we obtain that

$$\frac{\det H^{1/2}}{\det K} = \frac{(\mu_1 \cdots \mu_n)^{1/2}}{\lambda_1 \cdots \lambda_n} \leq \frac{\mu_{3k}^{3k/2} \cdot 1^{k/2}}{\lambda_1^k \cdot \varepsilon^{(n-k)}} \leq O(1).$$

The theorem then follows from Theorem 3.16.   □

4

Gromov's Systolic Inequality via Smoothing

Lizhi Chen[*]

A central theorem shown by Gromov in systolic geometry concerns the relation between two topological invariants: systolic volume and simplicial volume. The main method used to show this theorem is Gromov's smoothing technique, which relies on an alternative definition of simplicial volume. In this chapter, a brief introduction is given for the theorem and the smoothing technique.

4.1 Gromov's Systolic Inequality

Let M be a closed manifold of dimension n endowed with a Riemannian metric $\mathcal{G}$, denoted $(M, \mathcal{G})$. The *systole* of $(M, \mathcal{G})$, denoted by $\mathsf{Sys}\pi_1(M, \mathcal{G})$, is defined to be the shortest length of a non-contractible loop. A closed connected n-manifold M is *essential* if there exists a continuous map $f : M \to K$ from M to an aspherical topological space K, such that $f_n([M]) \neq 0$ in $\mathrm{H}_n(K; R)$, where $[M] \in \mathrm{H}_n(M; R)$ is the fundamental class of M, and the coefficient ring is $R = \mathbb{Z}$ if M is orientable and $R = \mathbb{Z}_2$ if M is non-orientable. Examples of essential manifolds include all closed hyperbolic n-manifolds, the n-dimensional torus $\mathbb{T}^n$, and the real projective n-space $\mathbb{R}P^n$. Moreover, all connected closed aspherical n-dimensional manifolds are essential, and connected sums of aspherical manifolds are also essential. The main theorem in the area of systolic geometry is the following systolic inequality proved by Gromov:

Theorem 4.1 [Gromov, 1983] *For any Riemannian metric $\mathcal{G}$ defined on a closed essential manifold M of dimension n,*

$$C_n \mathsf{Sys}\,\pi_1(M, \mathcal{G})^n \leqslant \mathrm{vol}_{\mathcal{G}}(M),$$

where C_n is a positive constant only depending on n.

* Supported by the Young Scientists Fund of NSFC (Award No. 11901261).

Moreover, Gromov found that for a given essential manifold M, the optimal constant in the systolic inequality is related to the topological complexity of M. In particular, there is a relation with simplicial volume. The optimal constant in the systolic inequality is usually called "systolic volume."

Definition 4.2 The *systolic volume* of a closed manifold M of dimension n, denoted $\mathsf{SR}(M)$, is defined to be

$$\inf_{\mathcal{G}} \frac{\mathrm{vol}_{\mathcal{G}}(M)}{\mathsf{Sys}\,\pi_1(M,\mathcal{G})^n},$$

where the infimum is taken over all Riemannian metrics $\mathcal{G}$ on M.

According to Theorem 4.1, systolic volume is positive for all essential manifolds. Babenko further showed [1992] that systolic volume is a homotopy invariant. Brunnbauer [2008] showed that systolic volume of connected closed manifolds only depends on the image of the fundamental class under the classifying map of the universal cover.

Denote by $\|M\|$ the simplicial volume of M. Gromov's theorem relating systolic volume to simplicial volume is given in the following:

Theorem 4.3 [Gromov, 1983, section 6.4.D'; Gromov, 1999 section 4.46; Gromov, 1996, section 3.C.3; see also Balacheff and Karam, 2019, section 4.3] *Assume that M is a closed manifold of dimension n with non-zero simplicial volume. The systolic volume $\mathsf{SR}(M)$ of M satisfies*

$$\mathsf{SR}(M) \geqslant D_n \frac{\|M\|}{\log^n \|M\|},$$

where D_n is a constant only depending on n.

The above theorem is a central theorem in systolic geometry, also considered to be the most difficult theorem in this area [Berger, 2003; Guth, 2010]. The proof mainly depends on the smoothing technique. It is our aim to give a brief introduction for this technique. We refer to Gromov [1982] and Balacheff and Karam [2019] for major references.

4.2 Straight Invariant Fundamental Cocycles

In the smoothing technique, we need to use straight invariant fundamental cocycles to give an alternative definition of simplicial volume.

Let M be a closed hyperbolic manifold of dimension n, and let hyp be the hyperbolic Riemannian metric defined on M. For a singular simplex $\sigma\colon \Delta^n \to M$, denote by $\tilde{\sigma}\colon \Delta^n \to \mathbb{H}^n$ the lifting of σ to the universal covering space $\mathbb{H}^n$. A *straight n-simplex* in the hyperbolic space $\mathbb{H}^n$ is defined inductively via the geodesic cone (see Section 2.3 for more details).

Note that a straight n-simplex in $\mathbb{H}^n$ is uniquely determined by its $(n+1)$ vertices. We denote by $\tilde{\sigma}_{st}$ the straight n-simplex with the same set of vertices of $\tilde{\sigma}$. Assume that $\pi\colon \mathbb{H}^n \to M$ is the universal covering map. Denote by σ_{st} the n-simplex $\pi \circ \tilde{\sigma}_{st}$. It is easy to see that σ_{st} and σ share the same set of vertices. Suppose $\omega \in \mathrm{H}^n(M;\mathbb{R})$ is a fundamental cocycle; then for a straight n-simplex σ_{st}, we have $\omega(\sigma_{st}) = \omega(\sigma)$. Therefore, the action of ω on an n-simplex σ is only determined by the set of vertices of σ. In this way, we can see ω as a real-valued function defined on M^n.

Proposition 4.4 *For the universal covering $\pi\colon \mathbb{H}^n \to M$, define the pullback $\pi^*\omega\colon \widetilde{M}^{n+1} \to \mathbb{R}$ as*

$$\pi^*\omega(y_0', y_1', \ldots, y_n') = \omega(\pi(y_0'), \pi(y_1)', \ldots, \pi(y_n')).$$

Then $\pi^*\omega\colon \widetilde{M}^{n+1} \to \mathbb{R}$ is also straight, that is, it only depends on the set of vertices.

Now let $(M, \mathcal{G})$ be any Riemannian manifold of dimension n, and $\pi\colon \widetilde{M} \to M$ be the universal covering map.

Definition 4.5 [Balacheff and Karam, 2019, definition 2.3]   A *straight invariant fundamental cocycle* is a cochain $\tilde{\omega} \in \mathrm{C}^n(\widetilde{M};\mathbb{R})$, such that

1. $\tilde{\omega}$ is $\pi_1(M)$-invariant;
2. the only cochain ω on M satisfying $\pi^*(\omega) = \tilde{\omega}$ is a cocycle representing the dual fundamental class of M, that is, $[\omega] = [M]^*$;
3. $\tilde{\omega}$ is straight, so that $\tilde{\omega}$ can be considered as a real valued function on $\widetilde{M}$ which only depends on $(n+1)$ vertices in $\widetilde{M}$; moreover, $\tilde{\omega}$ is Borel.

4.3 An Alternative Definition of Simplicial Volume

The standard definition of simplicial volume is given in terms of the fundamental class in the real coefficient homology group (see Definition 2.4 in Chapter 2, or Definition 5.1 in Chapter 5). There is also a dual definition of simplicial volume in terms of (bounded) cohomology [Balacheff and Karam, 2019, section 2.2]. In this section, we introduce an alternative definition of simplicial volume in terms of straight invariant fundamental cocycles. It can be shown that all these three definitions are equivalent.

For a given straight invariant fundamental cocycle $\tilde{\omega} \in C^n(\tilde{M}; \mathbb{R})$, its ℓ^∞-norm is defined to be

$$\|\tilde{\omega}\|_\infty = \sup |\tilde{\omega}(\tilde{y}_0, \tilde{y}_1, \ldots, \tilde{y}_n)|,$$

where the supremum is taken over all points $(\tilde{y}_0, \tilde{y}_1, \ldots, \tilde{y}_n) \in \tilde{M}^{n+1}$.

Definition 4.6 For the given Riemannian n-manifold $(M, \mathcal{G})$, define

$$\|M\|' = \frac{1}{\inf \|\tilde{\omega}\|_\infty},$$

where the infimum is taken over all straight invariant fundamental cocycles $\tilde{\omega}$ on $\tilde{M}$.

The above is an alternative definition of simplicial volume, due to Gromov [1982, section 2] and Balacheff–Karam [2019, section 2.3]. Following Gromov's ideas [Gromov, 1982, section 3], one can in fact prove that it is equivalent to the standard definition (Definition 1). For instance, in the hyperbolic case, we have the following:

Theorem 4.7 [Balacheff and Karam, 2019, theorem 2.4] *For a hyperbolic n-manifold M,*

$$v_n \|M\|' = \text{vol}_{hyp}(M),$$

where v_n is the maximal volume of an ideal regular n-simplex in $\mathbb{H}^n$ as in Section 2.1.

4.4 The Smoothing Technique

In the following, we still use $(M, \mathcal{G})$ to denote a closed Riemannian manifold of dimension n, and $\pi : \tilde{M} \to M$ to denote the universal covering.

Let $\mathcal{M}$ be the Banach space of all finite measures on $\tilde{M}$, and $\mathcal{P}$ be the subset of all probability measures. For $\mu \in \mathcal{M}$, its usual norm is

$$\|\mu\| = \int_{\tilde{M}} |\mu|,$$

where $|\mu|$ stands for the total variation. A straight invariant fundamental cocycle $\tilde{\omega}$ is extended to a real valued $(n+1)$-linear function on $\mathcal{M}^{n+1}$ as follows:

$$\tilde{\omega}(\mu_0, \mu_1, \cdots, \mu_n) = \int_{\widetilde{M}^{n+1}} \tilde{\omega}(\tilde{y}_0, \tilde{y}_1, \cdots, \tilde{y}_n) d\mu_0(\tilde{y}_0) d\mu_1(\tilde{y}_1) \cdots d\mu_n(\tilde{y}_n).$$

Definition 4.8 A *smoothing operator* is a smooth $\pi_1(M)$-equivariant map $\mathscr{S} \colon \widetilde{M} \to \mathcal{P}$.

For a straight invariant fundamental cocycle $\tilde{\omega}$ (seen as a real valued function on $\mathcal{M}^{n+1}$), the pullback $\mathscr{S}^*\tilde{\omega}$ is defined as follows:

$$\mathscr{S}^*\tilde{\omega}(\tilde{y}_0, \tilde{y}_1, \ldots, \tilde{y}_n) = \tilde{\omega}(\mathscr{S}(\tilde{y}_0), \mathscr{S}(\tilde{y}_1), \ldots, \mathscr{S}(\tilde{y}_n)).$$

The idea of the smoothing operator is to replace points in $\widetilde{M}$ by probability measures, and then observe the effect on diffused cochains.

Proposition 4.9 *Let $\mathscr{S} \colon \widetilde{M} \to \mathcal{P}$ be a smoothing operator and $\tilde{\omega}$ be a straight invariant fundamental cocycle. Then the pullback $\mathscr{S}^*(\tilde{\omega})$ is also a straight invariant fundamental cocycle.*

For a smoothing operator $\mathscr{S} \colon \mathcal{M} \to \mathcal{P}$, let $\tilde{y} \in \widetilde{M}$ and define

$$\|d\mathscr{S}_{\tilde{y}}\| = \sup_{\tau \in S} \|d\mathscr{S}_{\tilde{y}}(\tau)\|,$$

where the supremum is taken over all vectors τ in the unit tangent sphere $S_{\tilde{y}} \subset T_{\tilde{y}}\widetilde{M}$ of the pullback metric on the Riemannian universal covering $\widetilde{M}$. Then we define

$$\|d\mathscr{S}\|_\infty = \sup_{\tilde{y} \in \widetilde{M}} \|d\mathscr{S}_{\tilde{y}}\|,$$

where the supremum is over all $\tilde{y} \in \widetilde{M}$.

The smoothing inequality is given in the following theorem:

Theorem 4.10 (Gromov's smoothing inequality; see Balacheff and Karam [2019, theorem 3.4]) *Let $(M, \mathcal{G})$ be a closed Riemannian n-manifold. For a smoothing operator $\mathscr{S} \colon \widetilde{M} \to \mathcal{P}$ with $\|d\mathscr{S}\|_\infty < \infty$,*

$$\|M\| \leqslant n! \|d\mathscr{S}\|_\infty^n \operatorname{vol}_{\mathcal{G}}(M).$$

Remark 4.11 The proof of the above smoothing inequality uses the alternative definition $\|M\|'$ of simplicial volume. In Gromov's paper [Gromov, 1982], this part is contained in section 2.4.

4.5 Applications of the Smoothing Technique

The smoothing technique introduced above is applied to show Theorem 4.3. There are two main steps in the proof of Theorem 4.3. The first step concerns the existence of regular geometric cycles. The second step is the application of the smoothing technique. In Gromov's paper [Gromov, 1983], this theorem is contained in section 6.4. A detailed description of the existence of regular geometric cycles is in Bulteau [2015].

Let M be a closed Riemannian manifold of dimension n with $\|M\| > 0$. A geometric cycle (V, f) representing the fundamental class $[M]$ is a pseudomanifold V endowed with a piecewise Riemannian metric $\mathcal{G}$, such that the continuous map $f : V \to M$ satisfies $f_*([V]) = [M]$, where f_* stands for the induced homomorphism between top homology groups and $[V]$ stands for the fundamental class of V. A regular geometric cycle V^* satisfies $\mathsf{SR}(V^*) = \mathsf{SR}(M)$ and $\mathsf{Sys}\,\pi_1(V^*, \mathcal{G}) = 1$. Let $\widetilde{V^*}$ be the universal covering of V^*. When the smoothing technique is applied to show Theorem 4.3, the smoothing operator $\mathscr{S}(y)$ $(y \in \widetilde{V^*})$ constructed is defined to be a function $\mathscr{S} : \widetilde{V^*} \to \mathbb{R}$, so that

$$\mathscr{S}^* \mu(y') = \int_{V^*} \mathscr{S}(y, y') \mu(y') dy'.$$

With the chosen function $\mathscr{S}(y, y\prime)$, it is proved in Gromov [1983, section 6.4] that

$$\|d\mathscr{S}\|_\infty^n \leqslant \log^n \left(C_n \, \mathsf{SR}(V^*) \right),$$

where C_n is a positive constant only depending on n. More details can be found in Gromov [1982, section 2.4] and Gromov [1983, section 6.4]. Since the map $f : V^* \to M$ has degree one, $\|M\| \leqslant \|V^*\|$. Theorem 4.3 is proved by a combination with the above smoothing inequality.

More applications of the smoothing technique are mentioned in Balacheff and Karam's article [Balacheff and Karam, 2019].

5

Integral Foliated Simplicial Volume

Caterina Campagnolo[*]

The invariants considered hereafter are also defined for non-connected or non-orientable manifolds, or manifolds with boundary, but for simplicity we will always assume that M^n is a (non-empty) closed connected oriented manifold of dimension n.

5.1 A Question of Gromov

Recall the definition of simplicial volume (Definition 1):

Definition 5.1 The *simplicial volume* of M is

$$\|M\| = \inf\left\{\sum_{i=1}^{k} |a_i| \ \middle| \ \left[\sum_{i=1}^{k} a_i \sigma_i\right] = [M] \in \mathrm{H}_n(M; \mathbb{R})\right\} \in \mathbb{R}_{\geq 0},$$

where each $\sum_{i=1}^{k} a_i \sigma_i$ is a singular cycle with real coefficients, and $[M]$ is the fundamental class of M.

The following question was asked by Gromov:

Question 5.2 [Gromov, 1993] Let M be a closed connected oriented aspherical manifold. Then is it true that $\|M\| = 0$ implies $\chi(M) = 0$?

Recall that the Euler characteristic is defined by $\chi(M) = \sum_{j\geq 0}(-1)^j b_j(M)$, where $b_j(M)$ is the rank of $\mathrm{H}_j(M; \mathbb{Z})$. For later use, we note the equality $\chi(M) = \sum_{j\geq 0}(-1)^j b_j^{(2)}(\tilde{M})$, where $b_j^{(2)}(\tilde{M})$ is the von Neumann dimension of the ℓ^2-homology $\mathrm{H}_j^{(2)}(\tilde{M}; \mathbb{Z})$ of the $\pi_1(M)$-CW complex $\tilde{M}$, the universal covering of M with the deck transformation action of $\pi_1(M)$. See Chapter 6 for a definition and properties of ℓ^2-homology and ℓ^2-Betti numbers.

* Supported by the Swiss National Science Foundation, Postdoc.Mobility grant P400P2-191107/1.

Remark 5.3 Let us consider a few manifolds with vanishing simplicial volume for a quick sanity check of Gromov's question.

1. $\|S^2\| = 0$, $\chi(S^2) = 2$, but S^2 is not aspherical!
2. $\|T^n\| = 0$, $\chi(T^n) = 0$.
3. If M is a Euclidean manifold, it is finitely covered by a torus of the dimension of M. Then both the simplicial volume and the Euler characteristic of M vanish, by the preceding item and the multiplicativity of these invariants under finite coverings (Proposition 3).
4. If $\pi_1(M)$ is amenable, then $\|M\| = 0$. What about $\chi(M)$? The answer to this question will appear at the end of this chapter (see also Chapter 6, Section 6.3).

To approach Gromov's question above, and following the idea proposed by Gromov himself [1999], we will introduce a few new invariants related to the simplicial volume. We note here that there are some more strategies to approach the question: the reader is referred to the recent article by Löh, Moraschini, and Raptis for a state-of-the-art overview of the existing methods and new results [Löh et al., 2021].

5.2 Integral Simplicial Volume

Definition 5.4 The *integral simplicial volume* of M is

$$\|M\|_{\mathbb{Z}} = \inf\left\{\sum_{i=1}^k |a_i| \ \middle|\ \left[\sum_{i=1}^k a_i\sigma_i\right] = [M] \in H_n(M;\mathbb{Z})\right\} \in \mathbb{Z}_{\geq 1},$$

where each $\sum_{i=1}^k a_i\sigma_i$ is a singular cycle with integer coefficients, and $[M]$ is the fundamental class of M.

Remark 5.5 1. The only difference with Definition 5.1 is in the choice of the coefficients for the fundamental class of M and hence for the representing cycles.
2. With integer coefficients, the infimum becomes now a minimum.
3. $\|M\|_{\mathbb{Z}} \geq \|M\|$ as every integer cycle is also a real cycle.

The following lemma (see, for example, Frigerio et al. [2016, lemma 4.1], or Löh [2006b, proposition 4.1], Lück [2002, example 14.28] for slightly different versions) illustrates the strategy we will deploy to approach Gromov's question.

Lemma 5.6 $\|M\|_{\mathbb{Z}} \geq b_j(M)$ *for all* $j \geq 0$.

Proof Let $\sum_{i=1}^{k} a_i \sigma_i$ be a representative of $[M] \in H_n(M; \mathbb{Z})$ such that $\sum_{i=1}^{k} |a_i| = \|M\|_{\mathbb{Z}}$. Poincaré duality yields an isomorphism

$$H^{n-j}(M; \mathbb{Z}) \longrightarrow H_j(M; \mathbb{Z})$$

$$[\alpha] \longmapsto [\alpha] \cap [M] = \left[\sum_{i=1}^{k} a_i \alpha(\sigma_i \rfloor) \lfloor \sigma_i \right].$$

Then, for every $j \geq 0$, the module $H_j(M; \mathbb{Z})$ is a quotient of a module generated by at most k elements. Hence its rank $b_j(M)$ is at most k. Consequently,

$$b_j(M) \leq k = \sum_{i=1}^{k} 1 \leq \sum_{i=1}^{k} |a_i| = \|M\|_{\mathbb{Z}}. \qquad \square$$

This lemma inspires the following idea: the integral simplicial volume relates directly to the Betti numbers, and hence to the Euler characteristic. If it were equal to the simplicial volume for aspherical manifolds, the answer to Gromov's question 5.2 would be yes. Of course this is not the case, as the integral simplicial volume never vanishes. We will then try to find finer approximations of the simplicial volume, while keeping the same kind of inequalities as in Lemma 5.6. This motivates the definition of the invariants below.

5.3 Stable Integral Simplicial Volume

Definition 5.7 The *stable integral simplicial volume* of M is

$$\|M\|_{\mathbb{Z}}^{\infty} = \inf \left\{ \frac{\|N\|_{\mathbb{Z}}}{d} \mid N \longrightarrow M \text{ is a covering of degree } d \right\} \in \mathbb{R}_{\geq 0}.$$

Remark 5.8 1. We have $\|M\| \leq \|M\|_{\mathbb{Z}}^{\infty} \leq \|M\|_{\mathbb{Z}}$. The first inequality follows from the multiplicativity property of simplicial volume with respect to finite coverings (Proposition 3); the second inequality is obvious as the identity map is a covering of degree 1. For manifolds with trivial fundamental group, we have $\|M\|_{\mathbb{Z}}^{\infty} = \|M\|_{\mathbb{Z}}$.
2. Playing with coverings, it is easy to see that if $M' \to M$ is a covering of degree d', then $\|M'\|_{\mathbb{Z}}^{\infty} = d' \|M\|_{\mathbb{Z}}^{\infty}$.

Lemma 5.9 [Francaviglia et al., 2012, proposition 5.1]

$$(n+1)\|M\|_{\mathbb{Z}}^{\infty} \geq \chi(M).$$

Proof By Lemma 5.6, for every degree d' covering M' of M we have $\|M'\|_{\mathbb{Z}} \geq b_j(M')$ for every $j \geq 0$. Then

$$(n+1)\|M'\|_{\mathbb{Z}} \geq \sum_{j=0}^{n} |(-1)^j b_j(M')| \geq \chi(M') = d'\chi(M).$$

Dividing by d' and taking the infimum over all finite coverings of M yields the desired inequality. $\qquad\qquad\qquad\qquad\qquad\qquad\qquad\qquad\qquad\qquad\qquad\qquad\square$

The following result also holds true:

Lemma 5.10 $\|M\|_{\mathbb{Z}}^{\infty} \geq b_j^{(2)}(\widetilde{M})$ *for all* $j \geq 0$.

This can, for example, be deduced from Remark 5.16 (2) and Proposition 5.17 below.

In view of these lemmas, we get a new strategy to approach Gromov's question 5.2: if it is true that $\|M\|_{\mathbb{Z}}^{\infty} = \|M\|$, then the answer to the question is yes.

Let us consider the surface case:

Example 5.11 1. $\|S^2\|_{\mathbb{Z}}^{\infty} = \|S^2\|_{\mathbb{Z}} = 2$ because there are no non-trivial coverings of S^2, but $\|S^2\| = 0$.
2. $\|T^2\|_{\mathbb{Z}}^{\infty} = \|T^2\| = 0$ by multiplicativity under finite coverings.
3. $\|\Sigma_g\|_{\mathbb{Z}}^{\infty} = \|\Sigma_g\| = 4(g-1)$. Indeed, the surface Σ_h admits a triangulation by $4h-2$ triangles. Now one can construct an explicit degree d covering of Σ_g by Σ_h for every $d, g, h \in \mathbb{N}$ such that $2(1-h) = 2d(1-g)$ is satisfied. For such a degree d covering $\Sigma_h \to \Sigma_g$, we thus have the relation $h = d(g-1)+1$. Consequently,

$$\|\Sigma_g\|_{\mathbb{Z}}^{\infty} \leq \frac{\|\Sigma_h\|_{\mathbb{Z}}}{d} \leq \frac{4h-2}{d} = 4(g-1) + \frac{2}{d}$$

for every $d \in \mathbb{Z}_{>0}$. Thus, taking the infimum over d, we obtain

$$4(g-1) = \|\Sigma_g\| \leq \|\Sigma_g\|_{\mathbb{Z}}^{\infty} \leq 4(g-1)$$

and we have equality. This argument can be found in Gromov [1982, p. 9].

Other presently known cases of equality $\|M\| = \|M\|_{\mathbb{Z}}^{\infty}$ are listed hereafter. If M satisfies this equality, it is said to satisfy *integral approximation*:

1. M^3 hyperbolic [Frigerio et al., 2016];
2. M^3 aspherical [Fauser et al., 2021].

However, it was shown in Francaviglia et al. [2012] that for M^n hyperbolic, as soon as $n \geq 4$ we have $\|M^n\| < \|M^n\|_{\mathbb{Z}}^{\infty}$.

But recall from Theorem 2.1 that the simplicial volume of hyperbolic manifolds is positive. So the outlined strategy is still not lost for us, because the right question to ask is:

Question 5.12 If M is aspherical and $\|M\| = 0$, is it true that $\|M\|_{\mathbb{Z}}^{\infty} = 0$?

Remark 5.13 Often this question is asked with the additional assumption that $\pi_1(M)$ is residually finite, because it is unlikely that the answer is positive in full generality. However, we do not know any counterexample to the question formulated as above.

To Question 5.12, there are two more positive answers: manifolds with residually finite amenable fundamental group [Frigerio et al., 2016] and (generalised) graph manifolds with residually finite fundamental group [Fauser et al., 2019] satisfy integral approximation.

5.4 Integral Foliated Simplicial Volume

To define our next invariant, we need a bit more setup. Let $\pi: \widetilde{M} \to M$ be the universal cover of M. Denote by $\Gamma = \pi_1(M)$ the fundamental group of M. Then Γ acts on $\widetilde{M}$ by deck transformations. This induces a natural action of Γ on $C_{\bullet}(\widetilde{M}; \mathbb{Z})$. There is a canonical identification

$$C_{\bullet}(M; \mathbb{Z}) \cong \mathbb{Z} \otimes_{\mathbb{Z}\Gamma} C_{\bullet}(\widetilde{M}; \mathbb{Z}).$$

A *standard Borel space* is a measurable space that is isomorphic to a Polish space with its Borel σ-algebra $\mathcal{B}$. Recall that a *Polish space* is a separable, completely metrizable topological space. Let X be a *standard Borel probability space*, that is, a standard Borel space endowed with a probability measure μ. Suppose now that Γ acts (on the left) on such a standard Borel probability space $(X, \mathcal{B}, \mu)$ in a measurable and measure preserving way. Denote this action by $\alpha: \Gamma \to \mathrm{Aut}(X, \mu)$. Set

$$L^{\infty}(X, \mathbb{Z}) = \{f: X \longrightarrow \mathbb{Z} \,|\, \exists\, C \in \mathbb{R} : |f(x)| \leq C \text{ for } \mu\text{-a. e. } x \in X\}.$$

We define a right Γ-action on $L^{\infty}(X, \mathbb{Z})$ by setting

$$(f \cdot \gamma)(x) = f(\gamma x), \text{ for all } f \in L^{\infty}(X, \mathbb{Z}), \, \gamma \in \Gamma, \, x \in X.$$

There is a canonical inclusion for any $k \in \mathbb{Z}_{\geq 0}$:

$$i_{\alpha}: \quad C_k(M; \mathbb{Z}) \cong \mathbb{Z} \otimes_{\mathbb{Z}\Gamma} C_k(\widetilde{M}; \mathbb{Z}) \quad \longrightarrow \quad L^{\infty}(X, \mathbb{Z}) \otimes_{\mathbb{Z}\Gamma} C_k(\widetilde{M}; \mathbb{Z})$$
$$n \otimes \sigma \quad \longmapsto \quad \mathrm{const}_n \otimes \sigma,$$

where the tensor products are taken over the given actions. We will write $C_\bullet(M; \alpha)$ for the complex $L^\infty(X, \mathbb{Z}) \otimes_{\mathbb{Z}\Gamma} C_\bullet(\tilde{M}; \mathbb{Z})$. We call its elements *parametrized chains*.

Given a parametrized k-chain $z = \sum_i f_i \otimes \sigma_i \in L^\infty(X, \mathbb{Z}) \otimes_{\mathbb{Z}\Gamma} C_k(\tilde{M}; \mathbb{Z})$, we define its *parametrized ℓ^1-norm* as

$$|z|_1 = \sum_i \int_X |f_i| d\mu,$$

where each $\sigma_i : \Delta^k \to M$ is a singular simplex of dimension k. Here we assume that z is in reduced form, that is, if $i \neq j$, then $\pi \circ \sigma_i \neq \pi \circ \sigma_j$.

From now on we set:

$$H_\bullet(M; \alpha) := H_\bullet(C_\bullet(M; \alpha)).$$

Definition 5.14 [Gromov, 1999; Schmidt, 2005]   The *integral foliated simplicial volume* of M is the infimum over the parametrized ℓ^1-norms of the parametrized cycles representing the fundamental class:

$$|M| = \inf_{\alpha,(X,\mu)} \left\{ |M|^\alpha \mid \alpha : \Gamma \longrightarrow \mathrm{Aut}(X, \mu) \right\} \in \mathbb{R}_{\geq 0},$$

where the *α-parametrized simplicial volume* $|M|^\alpha$ is given by

$$\inf \left\{ \sum_i \int_X |f_i| d\mu \,\middle|\, [M]^\alpha = \left[\sum_i f_i \otimes \sigma_i \right] \in H_n(M; \alpha) \right\} \in \mathbb{R}_{\geq 0}.$$

Here $[M]^\alpha$ denotes the image of the fundamental class $[M] \in H_n(M; \mathbb{Z})$ in $H_n(M; \alpha)$ under the map induced by i_α.

Remark 5.15 Note that the infimum in the definition of $|M|$ is achieved by an *essentially free* action α [Löh and Pagliantini, 2016, corollary 4.14], meaning that $\mu(\{x \in X \mid Stab_\alpha(x) \neq \{e_\Gamma\}\}) = 0$.

Remark 5.16 1. $|M| \leq \|M\|_{\mathbb{Z}}$. Indeed, if α is the trivial action of Γ on $X = \{*\}$, then $|M|^\alpha = \|M\|_{\mathbb{Z}}$. Actually the equality holds as soon as α is trivial, even if X is not [Löh and Pagliantini, 2016, example 4.5]. So in particular $|M| = \|M\|_{\mathbb{Z}}$ if $\Gamma = \{1\}$ (see also Schmidt [2005, proposition 5.29]).
2. $|M| \leq \|M\|_{\mathbb{Z}}^\infty$. Indeed, if $X = \hat{\Gamma} = \lim_{\leftarrow N \trianglelefteq_{f.i.} \Gamma} \Gamma/N$ is the profinite completion of Γ, endowed with the pullback μ of the counting probability measures on the finite groups Γ/N, then (X, μ) is a standard Borel space, and the natural action α of Γ on X is measurable and measure preserving. Then $|M|^\alpha = \|M\|_{\mathbb{Z}}^\infty$ [Löh and Pagliantini, 2016, theorem 1.5]).

3. $\|M\| \leq |M|$. Indeed, consider

$$C_k(M; \mathbb{Z}) \longrightarrow L^\infty(X, \mathbb{Z}) \otimes_{\mathbb{Z}\Gamma} C_k(\widetilde{M}; \mathbb{Z}) \longrightarrow C_k(M; \mathbb{R})$$
$$\sum_i f_i \otimes \sigma_i \longmapsto \sum_i (\int_X f_i d\mu)\pi \circ \sigma_i.$$

Notice that the right map is norm non-increasing, and that in homology it maps a parametrized fundamental class of M to a real fundamental class of M. This shows the claim (see Schmidt [2005, theorem 5.35] or Löh and Pagliantini [2016, proposition 4.6]).

For more details on the integral foliated simplicial volume, see for example Löh and Pagliantini [2016].

Analogously to Lemma 5.6, it holds:

Proposition 5.17 $|M| \geq b_j^{(2)}(\widetilde{M})$ *for all* $j \geq 0$.

The idea of this statement goes back to Gromov [1999]. A proof can be found in Schmidt [2005, corollary 5.28] (the multiplicative constant therein can be improved to 1, to match the statement of Proposition 5.17).

So we are again in the position to propose a strategy: if $\|M\| = |M|$ for M aspherical, then the answer to Gromov's question 5.2 is yes (see also Chapter 6, Section 6.3). In fact, the definition of the integral foliated simplicial volume was proposed by Gromov precisely as a tool to study his question, and was formalized by Schmidt in his PhD thesis [Schmidt, 2005].

Let us again consider the surface case:

Example 5.18 1. $|S^2| = \|Z\|_{\mathbb{Z}} = 2$, as $\pi_1(S^2) = \{1\}$.
2. $|\Sigma_g| = \|\Sigma_g\|$ for every $g \geq 1$, as we have the sandwich inequality $\|\Sigma_g\| \leq |\Sigma_g| \leq \|\Sigma_g\|_{\mathbb{Z}}^\infty$, which is an equality by Example 5.11.

Presently it is known that we also have equality $\|M\| = |M|$ if M^3 is hyperbolic [Löh and Pagliantini, 2016, theorem 1.1]. However, as before, if M^n is hyperbolic, as soon as $n \geq 4$, we have $\|M^n\| < |M^n|$ [Frigerio et al., 2016, theorem 1.8].

But again, recall from Theorem 2.1 that the simplicial volume of hyperbolic manifolds is positive. So the outlined strategy is still not lost for us, because the right question to ask is:

Question 5.19 If M is aspherical and $\|M\| = 0$, is it true that $|M| = 0$?

To this question, there presently are a few positive answers:

1. if $\pi_1(M)$ is amenable and the dimension of M is non-zero [Frigerio et al., 2016, theorem 1.9] (which answers the question raised in Remark 5.3 (4));
2. if M is smooth and has trivial minimal volume [Braun, 2018, proof of corollary 5.4];
3. if M is smooth and admits a smooth non-trivial action by S^1 [Gromov, 1982; Yano, 1982; Fauser, 2021];
4. if M is smooth and admits a smooth regular foliation by S^1, with some technical assumptions [Campagnolo and Corro, 2021].

We are hoping for more, or for new approaches to Question 5.2!

Remark 5.20 The proof of Sauer [2009, theorem B] uses the *mass*, an invariant similar to the integral foliated simplicial volume: the main difference being that the associated norm only takes into account the measure of the support of the occurring functions, and not their values. He shows that if a triangulated aspherical manifold of dimension n is covered by amenable open sets with intersection multiplicity at most n, then its mass vanishes. He uses this result to show that in turn the ℓ^2-Betti numbers of this manifold vanish [Sauer, 2009, theorem A.1].

6

ℓ^2-Betti Numbers

Holger Kammeyer[*]

Typical spaces of interest in the study of simplicial volume and bounded cohomology are closed hyperbolic manifolds or more generally locally symmetric spaces. These are orbit spaces $M = X/\Gamma$ of a group action $\Gamma \curvearrowright X$ on a contractible space. To wit, the fundamental group $\Gamma = \pi_1 M$ acts by deck transformations on the globally symmetric universal covering $X = \widetilde{M}$. Strong rigidity results imply that not only the homotopy type but actually the isometry type of M is determined by the group Γ. So it stands to reason to put the group Γ into the focus of any thorough investigation of M and study the Γ-space X via equivariant methods. However, transferring the concepts of classical algebraic topology into an equivariant setting runs quickly into technical difficulties because X is not compact and Γ is an infinite group. In fact, functional analytic tools are necessary to cope with the infinite setting. The purpose of this chapter is to introduce the reader to the most established invariant that emerges in this way: the ℓ^2-*Betti numbers*.

A detailed treatment of the contents of this chapter can be found, for example, in Kammeyer [2019, chapters 2 to 6].

6.1 The Definition of ℓ^2-Betti Numbers

Let Γ be a group and let X be a Γ-*CW-complex*, by which we mean a CW-complex X with a left action of Γ by cellular homeomorphisms. Let us moreover assume that Γ acts freely and cocompactly on all the skeleta X_n of X. This has the effect that for each n, the set of open n-cells in X decomposes into a union of k_n simple transitive Γ-orbits. Correspondingly, the nth cellular chain complex $C_n(X) = H_n(X_n, X_{n-1})$, defined as the nth

* The author acknowledges financial support from Deutsche Forschungsgemeinschaft (German Research Foundation) grant no. KA 4641/2-1.

singular homology of the n-skeleton relative to the $(n-1)$-skeleton of X, carries the structure of a free left $\mathbb{Z}\Gamma$-module of rank k_n over the group ring $\mathbb{Z}\Gamma$. The differentials $\partial_n\colon C_n(X) \to C_{n-1}(X)$ in the cellular chain complex are by definition the boundary homomorphisms in the long exact homology sequence of the triple (X_n, X_{n-1}, X_{n-2}) and hence are Γ-equivariant by naturality. It follows that $C_\bullet(X)$ is a chain complex of finite rank free left $\mathbb{Z}\Gamma$-modules.

Definition 6.1 The ℓ^2-*chain complex* of X is the ℓ^2-completion

$$C_\bullet^{(2)}(X) := \ell^2\Gamma \otimes_{\mathbb{Z}\Gamma} C_\bullet(X).$$

Here we consider $\ell^2\Gamma$, the Hilbert space of square integrable complex valued functions on Γ, as a $\mathbb{C}\Gamma$-$\mathbb{Z}\Gamma$-bimodule. Choosing bases of $C_\bullet(X)$ in each degree, we obtain an isomorphism $C_n^{(2)}(X) \cong (\ell^2\Gamma)^{k_n}$ from which we transport a well-defined inner product to $C_n^{(2)}(X)$. Hence $C_n^{(2)}(X)$ is a Hilbert space with a left Γ-action by unitaries, and the ℓ^2-differentials $d_n^{(2)} = \mathrm{id}_{\ell^2\Gamma} \otimes \partial_n$ are bounded Γ-equivariant operators since they can be realised by right multiplication with matrices over $\mathbb{Z}\Gamma$ after choosing bases of $C_\bullet(X)$.

Definition 6.2 The nth ℓ^2-homology of X is given by

$$H_n^{(2)}(X) := \ker d_n^{(2)} / \overline{\mathrm{im}\, d_{n+1}^{(2)}}.$$

Here the bar denotes the closure in the Hilbert space. The quotient $H_n^{(2)}(X)$ can be canonically identified with the orthogonal complement of $\mathrm{im}\, d_{n+1}^{(2)}$ in $\ker d_n^{(2)}$ and thus is (isomorphic to) a closed Γ-invariant subspace of $(\ell^2\Gamma)^{k_n}$. Of course, we now want to define the nth ℓ^2-Betti number of X as the "equivariant dimension" of that subspace, pending a useful definition of "equivariant dimension." Instead of "equivariant dimension," it would be enough to have an "equivariant trace" available because we can then simply define the dimension of an invariant subspace as the trace of the orthogonal projection onto that subspace. Moreover, once we defined a trace on equivariant bounded operators on $\ell^2\Gamma$, we can extend it diagonally to equivariant operators on $(\ell^2\Gamma)^k$ for any k. The key to defining the trace of Γ-operators on $\ell^2\Gamma$ is the observation that $\ell^2\Gamma$ comes with a canonical *cyclic vector* for the action of Γ, namely the function $\delta \in \ell^2\Gamma$ with value one on the unit element $1 \in \Gamma$ and zero on any non-trivial $g \in \Gamma$. We define the trace as the corresponding matrix coefficient.

Definition 6.3 The *von Neumann trace* of a bounded Γ-operator T on $\ell^2\Gamma$ is

$$\mathrm{tr}_\Gamma(T) := \langle \delta, T\delta \rangle.$$

We agree that the inner product is anti-linear in the first and linear in the second variable. The trace property $\mathrm{tr}_\Gamma(TS) = \mathrm{tr}_\Gamma(ST)$ is easily verified if T and S lie in the image $\rho(\mathbb{C}\Gamma)$ of the right regular representation ρ of $\mathbb{C}\Gamma$ on $\ell^2\Gamma$, and the trace property remains true by continuity on the *weak closure* $\mathcal{R}(\Gamma)$ of $\rho(\mathbb{C}\Gamma)$ in the bounded operators $B(\ell^2\Gamma)$. By von Neumann's *bicommutant theorem* [Arveson, 1976, theorem 1.2.1] or Kammeyer [2019, theorem 2.19], $\mathcal{R}(\Gamma)$ can equivalently be defined as the bicommutant (commutant of the commutant) $\mathcal{R}(\Gamma) = (\rho(\mathbb{C}\Gamma))''$. Such weakly closed unital $*$-algebras of bounded operators are also known as *von Neumann algebras*, so the trace tr_Γ is in fact defined on the *group von Neumann algebra* $\mathcal{R}(\Gamma)$ – hence the name. A characteristic property of von Neumann algebras is that they contain many orthogonal projections. In particular, the group von Neumann algebra $\mathcal{R}(\Gamma)$ contains all orthogonal projections pr_U onto closed Γ-invariant subspaces $U \subset \ell^2\Gamma$.

Definition 6.4 The *von Neumann dimension* of a closed Γ-invariant subspace $U \subset (\ell^2\Gamma)^k$ is given by

$$\dim_\Gamma U = \mathrm{tr}_\Gamma \, \mathrm{pr}_U \in [0, k].$$

Now we are finally in shape to define ℓ^2-Betti numbers of a free Γ-CW-complex X with cocompact skeleta as von Neumann dimension of ℓ^2-homology.

Definition 6.5 The *nth ℓ^2-Betti number* of X is

$$b_n^{(2)}(X) := \dim_\Gamma \mathrm{H}_n^{(2)}(X).$$

6.2 Some Properties of ℓ^2-Betti Numbers

ℓ^2-Betti numbers have a variety of remarkable properties of which we shall present only three in this section to give you an indication of their strength and usefulness.

ℓ^2-Betti numbers are homotopy invariants. If two Γ-CW-complexes X and Y as above are Γ-homotopy equivalent, then $b_n^{(2)}(X) = b_n^{(2)}(Y)$ for all $n \geq 0$. This is more or less apparent because a Γ-homotopy equivalence can

be made cellular by cellular approximation; hence it induces a chain homotopy equivalence $C_\bullet(X) \simeq C_\bullet(Y)$ that we can complete to a chain homotopy equivalence $C_\bullet^{(2)}(X) \sim C_\bullet^{(2)}(Y)$. As an outcome, we can define ℓ^2-Betti numbers of groups. Recall that a *classifying space* $E\Gamma$ of a group Γ is a contractible free Γ-CW-complex. Classifying spaces always exist and are unique up to Γ-homotopy equivalence. If Γ is of *type* F_∞, meaning there exists a model for $E\Gamma$ with cocompact skeleta, we can thus agree on the following definition.

Definition 6.6 Suppose the group Γ has a classifying space $E\Gamma$ with cocompact skeleta. Then the *nth ℓ^2-Betti number* of Γ is given by

$$b_n^{(2)}(\Gamma) = b_n^{(2)}(E\Gamma).$$

Accepting a somewhat technical detour, one can define ℓ^2-Betti numbers of groups without the type F_∞ assumption, but actually many geometrically relevant groups have type F_∞, including fundamental groups of finite volume hyperbolic manifolds or, more generally, of finite volume locally symmetric spaces. Another useful remark is that one can equivalently define $b_n^{(2)}(\Gamma) := b^{(2)}(\underline{E\Gamma})$ where $\underline{E\Gamma}$ denotes the *classifying space for proper actions*. Typically, there exist simpler models for $\underline{E\Gamma}$ than for $E\Gamma$ [Lück, 2005].

Euler characteristic. The Euler characteristic of a finite CW-complex can be computed by the alternating sum of ordinary Betti numbers. Virtually the same proof shows that it can also be computed by the alternating sum of ℓ^2-Betti numbers of the universal covering. In our setting, this means

$$\chi(\Gamma \backslash X) = \sum_{n \geq 0} (-1)^n b_n^{(2)}(X) \tag{6.1}$$

for every free cocompact Γ-CW-complex X. So if the easily computed Euler characteristic (just count the cells) is non-zero, this detects that there exists non-trivial ℓ^2-homology in some degree, which is often not so easily established directly.

Lück approximation. Apart from the Euler characteristic identity, there is another more subtle asymptotic relationship between ℓ^2-Betti numbers $b_n^{(2)}(\cdot)$ and ordinary Betti numbers $b_n(\cdot)$. Suppose the countable group Γ is *residually finite*, meaning there exists a *residual chain* $\Gamma = \Gamma_0 \geq \Gamma_1 \geq \Gamma_2 \geq \cdots$ consisting of finite-index normal subgroups $\Gamma_i \trianglelefteq \Gamma$ such that $\bigcap_i \Gamma_i = \{1\}$. Again we consider a free Γ-CW-complex X with cocompact skeleta. One can consider the finite group actions $\Gamma/\Gamma_i \curvearrowright \Gamma_i \backslash X$ as an approximation of the action $\Gamma \curvearrowright X$. It is not hard to see that for these finite group actions, we have $b_n^{(2)}(\Gamma_i \backslash X) = \frac{b_n(\Gamma_i \backslash X)}{[\Gamma:\Gamma_i]}$. One may hope that ℓ^2-Betti numbers are "continuous" with respect to such an approximation, and Lück's approximation theorem reveals that this is indeed the case.

Theorem 6.7 [Lück, 1994] *We have*

$$b_n^{(2)}(X) = \lim_{i \to \infty} \frac{b_n(\Gamma_i \backslash X)}{[\Gamma : \Gamma_i]}.$$

So, remarkably, the limit is independent of the chosen residual chain. Existence of a residual chain is not guaranteed and in fact is equivalent to the residual finiteness of the group (see above). Fundamental groups of 3-manifolds are residually finite [Hempel, 1987; Agol, 2013], and so are finitely generated linear groups, which includes lattices in semisimple Lie groups with a finite center [Malcev, 1940]. Of course, every infinite simple group, for example a Burger–Mozes group [Burger and Mozes, 2000], is a non-example.

6.3 Relevance of ℓ^2-Betti Numbers

ℓ^2-Betti numbers have proven to be relevant in numerous contexts of which we shall only name three, corresponding to the three properties presented above.

ME proportionality. Though we have introduced ℓ^2-Betti numbers by means of an algebraic topology definition, they exhibit a surprising proportionality property from the viewpoint of measured group theory.

Theorem 6.8 [Gaboriau, 2002] *Let G be a locally compact group with Haar measure μ and let Γ, $\Lambda \le G$ be lattices. Then for all $n \ge 0$, we have*

$$\frac{b_n^{(2)}(\Gamma)}{\mu(G/\Gamma)} = \frac{b_n^{(2)}(\Lambda)}{\mu(G/\Lambda)}.$$

Here the Haar measure μ induces a G-invariant measure on G/Γ and G/Λ. Already the immediate corollary that lattices in the same locally compact group have vanishing or non-vanishing ℓ^2-Betti numbers in the same degrees could not really be anticipated because cocompact and non-cocompact lattices in the same group G behave quite differently in many other aspects. In fact, Gaboriau showed more generally that the ℓ^2-Betti numbers of *measure equivalent* groups obey a similar proportionality law. Since all infinite *amenable* groups form a single measure equivalence class by a famous theorem of Ornstein–Weiss [1980], they all have vanishing ℓ^2-homology because $b_n^{(2)}(\mathbb{Z}) = 0$ for all $n \ge 0$. This result was previously obtained by Cheeger and Gromov [1986]. So ℓ^2-Betti numbers can potentially be used to detect non-amenability of groups.

Euler characteristic and simplicial volume. Since we have now defined all
the necessary tools, let us repeat that ℓ^2-Betti numbers provide a possible proof
strategy for Gromov's question 5.2. The question asks whether for a closed con-
nected oriented aspherical manifold M, vanishing simplicial volume $\|M\| = 0$
implies vanishing Euler characteristic $\chi(M) = 0$. By Proposition 5.17, the
integral foliated simplicial volume $|M|$ satisfies $|M| \geq b_n^{(2)}(\widetilde{M})$ for all $n \geq 0$,
and by (6.1) we have $\chi(M) = 0$ if $b_n^{(2)}(\widetilde{M}) = 0$ for all $n \geq 0$. Hence:

If Question 5.19 has an affirmative answer, then so does Question 5.2.
See also Chapter 5, Section 5.4.

Homology growth. Lück's approximation theorem applied to the classifying
space of a type F_∞ residually finite group Γ shows that

$$b_n^{(2)}(\Gamma) = \lim_{i \to \infty} \frac{b_n(\Gamma_i)}{[\Gamma : \Gamma_i]}$$

for every residual chain $(\Gamma_i)_i$. This shows that if $b_n^{(2)}(\Gamma) > 0$, then the rank of
the free part of the homology $H_n(\Gamma_i; \mathbb{Z})$ must have linear asymptotic growth
with respect to the index $[\Gamma : \Gamma_i]$ and the asymptotic proportionality constant
is precisely $b_n^{(2)}(\Gamma)$. So ℓ^2-Betti numbers capture free homology growth along
residual chains. This raises the question of whether there might be another ℓ^2-
invariant that informs about torsion growth in homology, in particular in the
situation when all ℓ^2-Betti numbers vanish. For certain arithmetic groups Γ,
this question has been studied in detail by Bergeron–Venkatesh [Bergeron and
Venkatesh, 2013] for various coefficient systems and also for certain chains
of non-normal subgroups. A quick introduction to their work can be found
in Kammeyer [2019, section 6.6]. Conjecturally, exponential torsion growth
in homology should occur if and only if the ℓ^2-torsion $\rho^{(2)}(\Gamma)$ is positive.
Recent partial progress in answering these questions was obtained by Abert
et al. [2021] and Okun and Schreve [2021]. We refer the reader interested in the
general theory of ℓ^2-torsion to Lück [2002, chapter 3] and to Kammeyer [2019,
chapter 6], which emphasises the homology growth viewpoint. Interestingly,
integral foliated simplicial volume and hence also stable integral simplicial
volume is known to provide upper bounds for both free and torsion homology growth along general finite coverings (regular or not) of general closed
manifolds (aspherical or not), as is proven in Frigerio et al. [2016, theorem
1.6].

Part II

BOUNDED COHOMOLOGY

7

Stable Commutator Length

Nicolaus Heuer

What is stable commutator length? It is a real invariant which – just like any good group invariant – has several incarnations within mathematics. Indeed, I will give three different equivalent definitions of stable commutator length (from here on, scl), with some having a more topological, some a more algebraic, and some a more analytic flavor.

This allows us to interconnect several mathematical invariants and fields. We shall see, for example, how invariants from computer science, dynamical systems, and graph theory may be used to construct interesting simplicial volumes – via scl, of course.

The aim of this chapter is to give a curated overview over the different areas in stable commutator length, mostly due to personal taste.[1] The question "What is scl?" has been answered before by no less than Danny Calegari himself both in a short survey [Calegari, 2008] and in an extensive monograph [Calegari, 2009a]. The latter is the main reference for this introduction.

7.1 Three Ways to Stumble upon scl

There are three different categories through which one may stumble upon scl: algebraically, topologically, and analytically, yielding three different definitions of scl. In every case, we will use the definition to compute the same invariant: the stable commutator length of a commutator in the free group.

7.1.1 scl **via Commutators**

Let G be a group. Recall that a *commutator* is an expression $[x, y] = xyx^{-1}y^{-1}$ for $x, y \in G$. The group generated by all commutators is called the *commutator subgroup*, denoted by $[G, G]$. For an element

[1] And limited knowledge!

$g \in [G, G]$, the *commutator length* $\mathrm{cl}_G(g)$ measures how many commutators are needed to realise g as a product, that is,

$$\mathrm{cl}_G(g) := \min\{n \mid \exists x_i, y_i \in G : g = [x_1, y_1] \cdots [x_n, y_n]\}.$$

Definition 7.1 (scl algebraically) Let $g \in G$ be an element in a group. If $g \in [G, G]$ lies in the commutator subgroup, we define the *stable commutator length* of g in G as

$$\mathrm{scl}_G(g) := \lim_{n \to \infty} \frac{\mathrm{cl}_G(g^n)}{n}.$$

If some power g^N lies in the commutator subgroup, we set $\mathrm{scl}_G(g) := \frac{\mathrm{scl}_G(g^N)}{N}$. If no power of g lies in the commutator subgroup, we set $\mathrm{scl}_G(g) := \infty$.

Commutator length is easily seen to be subadditive and thus the defining limit exists.

Example 7.2 (Commutators in free groups) Let $F = F(\mathrm{a}, \mathrm{b})$ be the free group with free generating set $\{\mathrm{a}, \mathrm{b}\}$. Then $\mathrm{cl}_F([\mathrm{a}, \mathrm{b}]) = 1$ (as $[\mathrm{a}, \mathrm{b}]$ is a non-trivial commutator), $\mathrm{cl}_F([\mathrm{a}, \mathrm{b}]^2) = 2$ (as $[\mathrm{a}, \mathrm{b}]^2$ is not a commutator), and $\mathrm{cl}_F([\mathrm{a}, \mathrm{b}]^3) = 2$ (as scl is surprising and interesting[2]).

More generally, Culler [1981] found that $\mathrm{cl}_F([\mathrm{a}, \mathrm{b}]^n) = \lfloor \frac{n}{2} \rfloor + 1$. Taking the limit, we see that

$$\mathrm{scl}_F([\mathrm{a}, \mathrm{b}]) = \frac{1}{2}.$$

7.1.2 scl **via Surfaces**

Let X be a topological space with a loop $\gamma : S^1 \to X$. A natural measure for the complexity of γ is the complexity of a surface Σ needed to *fill* γ.

So what do we mean by *filling* a loop? We mean that there is a map $\Phi : \Sigma \to X$ such that the boundary $\partial \Sigma$ maps to X and factors through γ via a map $\partial \Phi : \partial \Sigma \to S^1$, that is, such that the diagram

$$
\begin{array}{ccc}
\partial \Sigma & \overset{i}{\longrightarrow} & \Sigma \\
\downarrow{\scriptstyle \partial \Phi} & & \downarrow{\scriptstyle \Phi} \\
S^1 & \overset{\gamma}{\longrightarrow} & X
\end{array}
$$

---

[2] If you don't believe me, here is one way to see this: $[\mathrm{a}, \mathrm{b}]^3 = [\mathrm{abA}, \mathrm{BabA}^2][\mathrm{Bab}, \mathrm{b}^2]$, where $\mathrm{A} = \mathrm{a}^{-1}$ and $\mathrm{B} = \mathrm{b}^{-1}$.

commutes. The degree of $\partial \Phi$ will be denoted by $n(\Sigma, \Phi)$. We will call such surfaces *admissible*.

By *complexity* we mean – of course – the Euler characteristic of the surface, except that we only consider non-spherical components. That is, we define $\chi^-(\Sigma) := \sum_{i=1}^n \min\{0, \chi(\Sigma_i)\}$, where Σ_i are the connected components of Σ. Finally we define:

Definition 7.3 (scl topologically) Let $\gamma \colon S^1 \to X$ be a loop in a topological space X. If there is no admissible surface to γ, we set $\mathrm{scl}_X(\gamma) := \infty$. Else, we set

$$\mathrm{scl}_X \left(\sum \gamma \right) := \inf_\Sigma \frac{1}{2} \cdot \frac{-\chi^-(\Sigma)}{n(\Sigma, \Phi)},$$

where the infimum ranges over all admissible surfaces with non-zero degree $n(\Sigma, \Phi)$.

Example 7.4 Consider as a topological space X a torus with a disk removed. Let $\gamma \colon S^1 \to X$ be the boundary loop of that disk. Then, the space X itself is a surface Σ with boundary and the identity map has degree 1. Note that Σ has genus one and one boundary component and thus $\chi^-(\Sigma) = -1$. We thus estimate

$$\mathrm{scl}_X(\gamma) \leq \frac{1}{2} \cdot \frac{-\chi^-(\Sigma)}{n(\Sigma, id)} = \frac{1}{2}.$$

We note that the fundamental group of X is the free group on two generators a and b (e.g., the meridian and longitude) and that γ corresponds to the commutator [a, b] in the fundamental group.

7.1.3 scl **via Quasimorphisms**

Let G be a group and let $g \in G$ be an element. One of the few objects one may compute (in the Turing sense) from a given presentation are its homomorphisms $G \to \mathbb{R}$. This, however, is rather limiting: We are not able to *see* any element $g \in [G, G]$ in its abelian quotients.

We could of course ask for different target groups, for example by studying G via its finite quotients. We will take a slightly different approach and instead generalize the type of morphisms to $\mathbb{R}$. This is one way to motivate the following:

Definition 7.5 (Quasimorphisms) A *quasimorphism* is a map $\phi \colon G \to \mathbb{R}$ such that there exists a constant D, such that $|\phi(g) + \phi(h) - \phi(gh)| \leq D$ for all $g, h \in G$. The infimum of all such D is called the *defect* of ϕ and denoted by $D(\phi)$.

Observe that a map ϕ is a quasimorphism with $D(\phi) = 0$ if and only if it is a homomorphism. Quasimorphisms form a vector space under pointwise addition and scalar multiplication. The space of such quasimorphisms is enormous: Indeed, any bounded function and any homomorphism is a quasimorphism. Quasimorphisms which lie in the vectorspace spanned by those maps will be called *trivial quasimorphisms*. We may get rid of bounded maps by only considering *homogeneous* quasimorphisms, that is, those quasimorphisms which additionally satisfy that $\phi(g^n) = n \cdot \phi(g)$ for all $n \in \mathbb{Z}$ and $g \in G$. It may be seen[3] that every quasimorphism has a unique homogeneous quasimorphism in bounded distance. Homogeneous quasimorphisms have some interesting properties, for example $\phi(g) = \phi(hgh^{-1})$ for all $g, h \in G$ [Calegari, 2009a, 2.2.3].

Fix an element $g \in [G, G]$. We may see that for any quasimorphism ϕ, we may *bound* $\phi(g)$ uniformly in terms of $D(\phi)$ and $\mathrm{cl}(g)$: For example, for any commutator $[x, y]$ we can write $|\phi([x, y])| \leq |\phi(x) + \phi(y) + \phi(x^{-1}) + \phi(y^{-1})| + 3D(\phi) \leq 5D(\phi)$ by successively applying the definition of quasimorphisms. On the other hand, if $g \in G$ is such that there is some homomorphism $\phi\colon G \to \mathbb{R}$ with $\phi(g) > 0$, then we may never bound $\phi(g)$ just in terms of $D(\phi)$, since we may arbitrarily scale up ϕ. We thus consider the following:

Definition 7.6 (scl via quasimorphisms; [Bavard, 1991])   Let G be a group and let $g \in G$ be an element. Then

$$\mathrm{scl}_G^{qm}(g) := \sup_\phi \frac{1}{2} \cdot \frac{\phi(g)}{D(\phi)},$$

where the supremum ranges over all homogeneous quasimorphisms $\phi\colon G \to \mathbb{R}$ with defect $D(\phi)$.

Example 7.7 (Brooks quasimorphisms) Let $F = F(a, b)$ be the non-abelian free group with free generating set $\{a, b\}$. We will also write A for a^{-1} and B for b^{-1}. Fix a word $x \in F$. We denote by $\nu_x\colon F \to \mathbb{N}$ the map that associates to a word w the largest number of times x is a subword of w, that is, the maximum of all n such that

$$w = w_0 \cdot x \cdot w_1 \cdots w_{n-1} \cdot x \cdot w_n$$

for appropriate w_i and where this expression is reduced. We define $\phi_x\colon F \to \mathbb{Z}$ via $\phi_x(w) \mapsto \nu_x(w) - \nu_{x^{-1}}(w)$. It turns out that ϕ_x is a quasimorphism with

---

[3] Consider for a quasimorphism $\phi\colon G \to \mathbb{R}$ the map $\bar\phi(g) := \lim_{n\to\infty} \frac{\phi(g^n)}{n}$, called the *homogenization of ϕ*. We have that $\bar\phi$ is a quasimorphism and $D(\bar\phi) \leq 2 \cdot D(\phi)$ [Calegari, 2009a, lemma 2.21].

$D(\phi_x) \leq 3$, called *Brooks quasimorphism*. Those maps were originally introduced by Brooks [1981] to show that the space of quasimorphisms on the free group is infinite-dimensional.

Consider the element $[a, b] \in F$. It may be seen [Heuer, 2019c, section 2.4] that $\phi := \phi_{ab} - \phi_{ba}$ satisfies $D(\phi) = 1$ and the associated homogeneous quasimorphism $\bar{\phi}$ satisfies $D(\bar{\phi}) = 2$. Moreover one may compute that $\bar{\phi}([a, b]) = 2$.

Putting things together, we estimate

$$\mathrm{scl}_F^{qm}([a, b]) \geq \frac{1}{2} \cdot \frac{\bar{\phi}([a, b])}{D(\bar{\phi})} = \frac{1}{2}.$$

Quasimorphisms (and thus stable commutator length) are directly related to bounded cohomology as follows (see Chapter 8 for more details):

Proposition 7.8 [Calegari, 2009a, theorem 2.50] *Let G be a group. The vector space of quasimorphisms $G \to \mathbb{R}$ modulo the vector space of trivial quasimorphisms is in a 1-1 correspondence to $\ker(c^2)$, the kernel of the comparison map $c^2 \colon \mathrm{H}_b^2(G; \mathbb{R}) \to \mathrm{H}^2(G; \mathbb{R})$.*

7.1.4 Wrapping Things Up: Equivalence of Definitions

Of course, all of the Definitions 7.1, 7.3, and 7.6 of stable commutator length are equivalent!

Theorem 7.9 [Calegari, 2009a, proposition 2.10; Bavard, 1991] *Let X be a topological space with fundamental group G and let $\gamma \colon S^1 \to X$ be a loop corresponding to the element $g \in G$. Then*

$$\mathrm{scl}_X(\gamma) = \mathrm{scl}_G(g) = \mathrm{scl}_G^{qm}(g).$$

Thus Examples 7.2, 7.4, and 7.7 computed and estimated exactly the same fact, namely that the stable commutator length of a commutator in the nonabelian free group on two generators is $\frac{1}{2}$. Note the different nature of the invariants involved: scl_X is an infimum, scl_G is a limit, and scl_G^{qm} is a supremum. It turns out that the infimum in scl_X is only sometimes achieved, while the supremum in scl_G^{qm} is always achieved. The quasimorphisms which achieve this supremum are also called *extremal quasimorphisms*.

Since all definitions are equivalent, we will only now study scl on groups. We also note that scl generalises to formal sums of elements, called *chains*.

We collect some basic properties:

Proposition 7.10 (Monotonicity [Calegari, 2009a, lemma 2.4])   *For any homomorphism* $\Phi: G \to H$ *between two groups, we have that* $\mathrm{scl}_G(g) \geq \mathrm{scl}_H(\Phi(g))$ *for all* $g \in G$. *From this we see that* scl *is invariant under automorphisms and invariant under retractions.*

Proposition 7.11 (Finite-index subgroups [Calegari, 2009a, corollary 2.81])   *If* $H < G$ *is a normal finite-index subgroup and* $h \in H$, *then we may compute* $\mathrm{scl}_H(h)$ *in terms of* scl_G *as follows:*

$$\mathrm{scl}_G(h) = \frac{1}{[G:H]} \cdot \mathrm{scl}_H \left(\sum_{aH \in G/H} aha^{-1} \right).$$

7.2 Vanishing, Gaps, and Lions

So what is stable commutator length? After having seen three equivalent definitions, we will explore scl on finitely presented groups.

Martin Bridson [2006, figure 1] charted the landscape of finitely presented groups by means of their large-scale geometry. Starting from $\mathbb{Z}$, one may explore finitely presented groups in two very different directions: One may follow the amenable path, passing by (in increasing difficulty) the abelian, nilpotent, polycyclic, and solvable groups. One may also wander off in the directions of groups of negative (or non-positive) curvature, passing by free, hyperbolic, semi-hyperbolic, CAT(0), and acylindrically hyperbolic groups.

The limits of exploring finitely presented groups are embodied by lions: as the workings of any Turing machine may be encoded in a finitely presented group, and as basic questions of Turing machines are undecidable, there is no hope of understanding or computing meaningful invariants from arbitrary finitely presented groups.

We will see that scl respects this landscape: either scl vanishes on the whole group (in the amenable case), or scl of a group may be uniformly bounded from below (for groups with non-positive curvature). And we will see that also scl cannot tame the lions, though there has been some progress by cornering them to *right-computable numbers* (Theorem 7.15).

7.2.1 Vanishing

Stable commutator length vanishes for any group with trivial real second bounded cohomology. This implies that $\mathrm{scl}_G(g) = 0$ for any amenable group

G and $g \in G$. This is a huge class of groups, encompassing, for example, all solvable groups. Besides this some other vanishing results are known, for example for subgroups of piecewise linear transformations of the interval [Calegari, 2007].

7.2.2 Gaps

In contrast, many classes of non-positively curved groups have a *gap* in stable commutator length. A group G is said to have such a *spectral gap* if there is a constant $C > 0$ such that for any element $g \in G$ either $\text{scl}_G(g) = 0$ or $\text{scl}_G(g) \geq C$. The largest C is called the *optimal gap* and denoted by C_G. Typically, one may also control the elements which satisfy $\text{scl}_G(g) = 0$, and we say that G has a *strong gap* if the only element satisfying $\text{scl}_G(g) = 0$ is the identity. We will see that all groups satisfy $C_G \leq \frac{1}{2}$ and that a gap of $\frac{1}{2}$ is obtained for a large class of non-positively-curved groups such as free groups and right-angled Artin groups.

Why might such a gap be useful? It allows us to obstruct and bound subgroups as follows: Suppose that $H < G$ is a subgroup of G and that G has a strong scl gap C_G. It follows from the monotonicity (Proposition 7.10) that H also has a strong scl gap and that $C_H \geq C_G$. Thus, a group G with a strong gap of $C_G = \frac{1}{2}$ only has subgroups with strong gaps $C_H = \frac{1}{2}$. This may be seen as some crude algebraic Tits alternative. Moreover, if $H < G$ is a finite-index subgroup, and H has a gap $\tilde{C}_H$ for *chains*, by the index formula (Proposition 7.11) we may estimate that $\tilde{C}_H \leq \frac{1}{[G:H]} \tilde{C}_G$. This allows one to estimate the indices of subgroups.

We list some notable results on scl gaps:

Theorem 7.12 *We have spectral gaps in the following cases:*

1. *Any Gromov hyperbolic group [Calegari and Fujiwara, 2010, theorem A], though this gap is not uniform. An element g has $\text{scl}_G(g) = 0$ if and only if g^n is conjugate to g^{-n} for some $n \in \mathbb{Z}_+$.*
2. *Any finite-index subgroup of the mapping class group $\text{MCG}(\Sigma)$ of a possibly punctured closed orientable surface Σ [Bestvina et al., 2016, theorem B]. There is a similar characterisation for elements with vanishing scl.*
3. *The fundamental group of any 3-manifold [Chen and Heuer, 2019, theorem C].*
4. *Any (subgroup of a) RAAG, in particular any special group, even has a strong gap of precisely $\frac{1}{2}$ [Heuer, 2019c]. See also Fernós et al. [2019] and Forester et al. [2020].*

5. *Elements in free and certain amalgamated free products which do not conjugate into vertex groups [Clay et al., 2016; Chen, 2018].*
6. *Elements in graph products which do not conjugate into vertex groups [Chen and Heuer, 2020].*

7.2.3 Lions

Beyond the realms of amenability and hyperbolicity lie the lions, the groups impossible to slay by means of Turing machines. Given a finitely presented group and an element $g \in G$, it is undecidable if $\mathrm{scl}_G(g) = 0$ or even if $\mathrm{scl}_G(g) \leq C$ for any positive real number C[4]. However, we may somewhat corner the lions (i.e., arbitrary finitely presented groups): we will see (Theorem 7.15) that the scls of finitely presented groups are always right-computable.

7.3 Spectrum

We now explore the spectrum of scl for a given group or class of groups, that is, the set $\mathrm{scl}_G(G) \subset \mathbb{R}_{\geq 0}$. We will start with the free group. Calegari [2009b] found an algorithm[5] to compute scl on free groups. The algorithm showed that scl is rational on free groups. The algorithm also allowed for computer experiments on the distribution of scl of random elements, which revealed a striking distribution; see Figure 7.1.[6] I emphasise that the only thing known about this figure is that scl $\geq 1/2$ and scl is rational – that's it! By merely looking at this figure, we may make two educated guesses (if not conjectures): The spectrum gets very sparse to the left, that is, there seems to be a second gap in scl[7], and scl seems to be much more frequent on elements with low denominator, – in particular the frequency of $p/(2q)$ seems to be proportional to q^{-d} for $d \sim 2$. A statistical analysis of this phenomenon may be found in Chen and Heuer [2020, section 7.3].

---

[4] Here is one way to see this: For a finitely presented group G and an element $g \in G$ we will construct an element $\tilde{g}$ in a group $\tilde{G}$ such that $\mathrm{scl}_{\tilde{G}}(\tilde{g}) = 0$ if and only if g is trivial in G and $\mathrm{scl}_{\tilde{G}}(\tilde{g}) = \frac{1}{2}$, else. We may assume that g is infinite torsion by replacing g with $g_1 g_2 \in G_1 \star G_2$ in $G_1 \star G_2$, where both g_i and G_i for $i \in \{1, 2\}$ are a copy of g and G, respectively. We may then observe that $\tilde{g} = [g, t] \in G \star \langle t \rangle := \tilde{G}$ is trivial if and only if g is trivial in G and $\mathrm{scl}_{\tilde{G}}(\tilde{g}) = \frac{1}{2}$, using the work of Chen [2018]. Thus, for any $C > 0$, we see that $\mathrm{scl}(\tilde{g}^{N(C)}) < C$ if and only if g is trivial, where $N(C) = 2 * \lceil C \rceil$. Thus computing scl is as undecidable as the word problem.
[5] By the name of `scallop`, which has been implemented by Alden Walker [Calegari and Walker, 2009]. It may be downloaded on Github, is very fast, and very interesting to play around with!
[6] The data set of the 50,000 random scls and the (Python) code to generate this figure may be found at www.nicolausheuer.com/code.html.
[7] Formally, there seems to be no $g \in F$ such that $1/2 < \mathrm{scl}_F(g) < 7/12$.

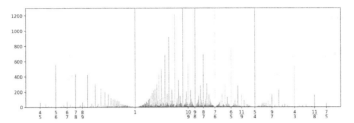

Figure 7.1 Histogram of scl of 50,000 random words of length 24 in $[F_2, F_2]$ using scallop [Calegari and Walker, 2009]

One may generalise the rationality of scl to certain free products [Chen, 2018] and to certain amalgamated free products and HNN extensions, including Baumslag–Solitar groups [Susse, 2015; Chen, 2020].

Besides free groups, very little is known about the spectrum of scl for other hyperbolic groups. It is unknown if scl is rational in surface groups, besides for certain elements [Forester and Malestein, 2020].

Which other values may scl take? A big source of examples comes from circle actions. There is a well-established connection between circle actions and bounded cohomology due to Ghys [2001]. Given a group G with an action on the circle $\rho\colon G \to \mathrm{Homeo}^+(S^1)$, this action allows us to cyclically extend G via the Euler class associated with ρ[8] to a group $\tilde{G}$. Then the action ρ on G lifts to an action $\tilde{\rho}\colon \tilde{G} \to \mathrm{Homeo}^+_{\mathbb{Z}}(\mathbb{R})$, the group of orientation preserving homomorphisms $\phi\colon \mathbb{R} \to \mathbb{R}$ of the reals which commute with the integers, that is, such that $\phi(x + n) = \phi(x) + n$ for all $x \in \mathbb{R}$ and $n \in \mathbb{Z}$. We define the *rotation number* of such ϕ via

$$\mathrm{rot}\colon \phi \longmapsto \lim_{n\to\infty} \frac{\phi^n(0)}{n}.$$

A key insight is that $\mathrm{rot}\colon \mathrm{Homeo}^+_{\mathbb{Z}}(\mathbb{R}) \to \mathbb{R}$ is a homogeneous quasimorphism of defect 1, which hence defines a quasimorphism on $\tilde{G}$ by pulling rot back via $\tilde{\rho}$. We have

Theorem 7.13 (scl and rotation number; [Calegari, 2009a, section 5]) *Let G be a perfect group that satisfies* $\mathrm{scl}_G(G) = 0$ *and that admits a non-trivial action* $\rho\colon G \to S^1$ *on the circle. Then for any element* $\tilde{g} \in \tilde{G}$ *in the central extension of G associated with the Euler class of ρ we have that*

$$\mathrm{scl}_{\tilde{G}}\colon \tilde{g} \longmapsto \frac{|\mathrm{rot}(\tilde{g})|}{2D},$$

---

[8] An introduction to this may be found in Bucher et al. [2016].

where rot *is the rotation number of* $\tilde{g}$ *and* D *is the defect of* rot *as a quasimorphism of* $\tilde{G}$.

Note that this already shows that $\mathrm{scl}_{\mathrm{Homeo}_{\mathbb{Z}}^{+}(\mathbb{R})}(\mathrm{Homeo}_{\mathbb{Z}}^{+}(\mathbb{R})) = \mathbb{R}_{\geq 0}$. Rotation number has been well studied for several groups acting on the circle. An interesting example of groups with vanishing scl is the group of piecewise linear transformations of the interval, as mentioned in Section 7.2.1.

For example, using Thompson's group T in Theorem 7.13, one constructs a finitely presented group whose scl spectrum is *exactly* $\mathbb{Q}_{\geq 0}$ [Calegari, 2009a, remark 5.20]. Zhuang used Stein–Thompson's groups to give the first example of finitely presented groups that have non-rational scl.

Theorem 7.14 [Zhuang, 2008] *There are finitely presented groups which have transcendental stable commutator lengths.*

All of the stable commutator lengths Zhuang constructed are a quotient of logarithms (e.g., $\log(3)/\log(2)$). Such numbers are either rational or transcendental. It is unknown if there are finitely presented groups that admit scls which are algebraic and not rational. I also mention that, using a connection to the fractional stability number of graphs, one may construct groups with exotic spectrum, such as groups that have a gap but are eventually dense [Chen and Heuer, 2020, theorem I, J].

More is known by considering *recursively presented* groups. These are all finitely generated subgroups of a finitely presented group. Note that the set of recursively presented groups is countable, and thus so is the set of scls on it. It is possible to characterize the set of scls on this class of groups by their computability.

Theorem 7.15 [Heuer, 2019b, theorem A] *The set of stable commutator lengths on recursively presented groups equals the set of non-negative right-computable numbers.*

A non-negative real number α is called right-computable if there is a Turing machine T which for any $i \in \mathbb{N}$ returns a rational number $T(i) \geq 0$ such that $T(i+1) \leq T(i)$ for all $i \in \mathbb{N}$ and $\alpha = \lim_{i \to \infty} T(i)$.

7.4 Relationship to Simplicial Volume

One application of stable commutator length is to construct manifolds with controlled simplicial volume.

Theorem 7.16 [Heuer and Löh, 2021a, theorem F]   *Let G be a finitely presented group such that $H_2(G; \mathbb{R}) \cong 0$. Then, for any $g \in [G, G]$, there is an orientable closed connected manifold M such that*

$$\|M\| = 48 \cdot \mathrm{scl}_G(g),$$

where $\|M\|$ denotes the simplicial volume of M.

We may use this theorem to translate the spectral results for scl known from Section 7.3 to the simplicial volume of manifolds. Using such techniques, we see

Theorem 7.17 [Heuer and Löh, 2020, 2021a,b]

1. *The set of simplicial volumes of oriented closed connected n-manifolds is dense in $\mathbb{R}_{\geq 0}$ for all $n \geq 4$.*
2. *Every rational is the simplicial volume of an oriented closed connected 4-manifold. Moreover, there is a sequence M_i of oriented closed connected 4-manifolds such that $\|M_i\| \to 0$ and such that $\|M_i\|$ are all linearly independent over the algebraic numbers and in particular transcendental.*
3. *The set of locally finite simplicial volumes of oriented connected open n-manifolds is $\mathbb{R}_{\geq 0}$ for any $n \geq 4$.*

7.5 Open Questions in scl

I end this chapter by listing some open questions about stable commutator length.

1. What are extremal quasimorphisms for arbitrary elements of the free group?
2. Is there a second gap of scl in non-abelian free groups F, that is, are there no elements $g \in F$ such that $\frac{1}{2} < \mathrm{scl}_F(g) < \frac{7}{12}$?
3. Is there a finitely presented group that has algebraic but not rational values of scl?[9] Is the set of scls on finitely presented groups the set of right-computable numbers?
4. Is scl rational on surface groups? If yes, is this rationality achieved using extremal surfaces? What about scl on Gromov hyperbolic groups?

    This is, at least qualitatively, related to Gromov's question: Does every one-ended hyperbolic group contain a surface subgroup [Gromov, 1987]?
5. In the free group, is there a connection to the primitivity rank in free groups? Recall that for an element $w \in F$ in a free group F the primitivity rank is

---

[9] This was recently answered positively [Fournier-Facio and Lodha, 2021].

defined as $\pi(w) = \min\{\mathrm{rk}(H)\}$, where $H < F$ runs over all subgroups of F such that $w \in H$ is *not* primitive in H. It was shown that the primitivity rank plays a crucial role in understanding the geometry of its associated one-relator subgroup [Louder and Wilton, 2018]. It was conjectured [Heuer, 2019a, conjecture 6.3.2] that for all $w \in F$, $\mathrm{scl}(w) \geq \frac{\pi(w)-1}{2}$. This would generalize the gap for elements in free groups. This conjecture has been verified for all words up to length 16 in free groups [Cashen and Hoffmann, 2020].

8

Quasimorphisms on Negatively Curved Groups

Biao Ma

As we have learned in Chapter 7, quasimorphisms on a group are homomorphisms from the group to the additive group of real numbers up to bounded errors. They play an important role for understanding both stable commutator length and second bounded cohomology. A central problem is to construct non-trivial quasimorphisms. It turns out that, by generalizing Brooks's construction for quasimorphisms on free groups, one can construct non-trivial quasimorphisms for many groups that share enough features of negative curvature. Examples of such groups include hyperbolic groups and mapping class groups. All such constructions are based on group actions on hyperbolic metric spaces, and they can be unified via introducing a geometric condition on actions, namely, *weak proper discontinuity (WPD)*.

In this chapter, we will talk about quasimorphisms on groups from the viewpoint of group actions. We will follow Bestvina and Fujiwara [2002], presented in a heuristic way.

8.1 Bounded Cohomology and Quasimorphisms

In this section, we give an introduction to (bounded) cohomology and quasimorphisms of the discrete group G. Our approach toward (bounded) group cohomology is via the *bar resolution* (Definition 14); since the homogeneous one will not appear, we denote it simply by $(C^\bullet_{(b)}(G, \mathbb{R}), \delta^\bullet)$. Recall that the inclusions $C^n(G) \to C^n_b(G)$ induce maps in cohomology $c^n \colon H^n_b(G) \to H^n(G)$, called *comparison maps*. When $n = 0$, the comparison map is the identity, and when $n = 1$ it is the zero map, so we only consider $n \geq 2$. The comparison maps c^n carry geometric information on the group G. For example, the comparison map $c^2 \colon H^2_b(G, V) \to H^2(G, V)$ is surjective for any bounded G-module V if and only if G is hyperbolic ([Mineyev, 2002]; this is Theorem 17 in the introduction).

We now focus on the kernel of the comparison map c^2, which was already mentioned in Proposition 7.8. We first recall the definition of quasimorphisms on a group. A *quasimorphism* on a group G is intuitively a homomorphism from G to $(\mathbb{R}, +)$, up to a bounded error. To be more precise:

Definition 8.1 Let G be a group; a function $\phi: G \rightarrow \mathbb{R}$ is called a *quasimorphism* if there exists a number $D(\phi) \geq 0$ such that, for any two elements g_1 and g_2 in G, we have $|\phi(g_1 g_2) - \phi(g_1) - \phi(g_2)| \leq D(\phi)$.

Let $\mathcal{L}(G)$ denote the space of all quasimorphisms on G. Since bounded functions are at a bounded distance from the trivial homomorphism, we are going to ignore them. Namely, if $BD(G)$ is the subspace of bounded functions on G, then we will consider the quotient space $QH(G)$ defined by $QH(G) = \mathcal{L}(G)/BD(G)$. Non-trivial elements in $QH(G)$ are called *non-trivial quasimorphisms* on G. There is also a subspace that is quite interesting but undetectable by the comparison map, namely $H^1(G) = \mathrm{Hom}(G, \mathbb{R})$. All facts mentioned above can be summarized in an exact sequence.

Proposition 8.2 [Frigerio, 2017, proposition 2.8] *The following sequence is exact:*

$$0 \longrightarrow H^1(G) \longrightarrow QH(G) \longrightarrow H^2_b(G) \xrightarrow{c^2} H^2(G).$$

Remark 8.3 Some authors define *trivial quasimorphisms* to be those that are at a bounded distance from a homomorphism. This is the case, for instance, in Calegari [2008], Frigerio [2017], and in Chapters 7 and 9. Since Bestvina and Fujiwara [2002] is the main reference for this chapter, we use their definition instead, according to which an unbounded homomorphism is a non-trivial quasimorphism. Note, however, that if G is finitely generated (or more generally if its abelianisation is finitely generated), then the space of homomorphisms $G \rightarrow \mathbb{R}$ is finite-dimensional. So if the space of non-trivial quasimorphisms of G is infinite-dimensional according to one definition, it is also infinite-dimensional according to the other.

We end this section with examples on groups with some results concerning comparison maps in degree 2.

Example 8.4 • [Johnson, 1972; Gromov, 1982] If G is amenable, then c^2 is injective. This is a special case of Theorem 18.
• [Burger and Monod, 1999] If G is a cocompact irreducible lattice in a semisimple Lie group of higher rank, then c^2 is injective.
• [Brooks, 1981] If G is a free group of rank at least 2, then c^2 is not injective. In fact, the kernel of c^2 is uncountably dimensional.

8.2 Hyperbolic Groups and Mapping Class Groups

As mentioned in the introduction, all constructions that we are going to talk about are based on group actions on hyperbolic spaces. We shortly recall some properties of hyperbolic metric spaces, hyperbolic groups, and mapping class groups of surfaces of finite type.

8.2.1 Hyperbolic Metric Spaces

A metric space (X, d_X) is called *geodesic* if for any two points p and q, there exists a geodesic γ in X, that is, an isometric embedding from the interval $[0, d_X(p, q)]$ of the real line to X, connecting p to q. Examples of geodesic metric spaces are given by connected graphs (not necessarily locally finite) where edges are assigned length one. For any two points p and q in X, we will use $[p, q]$ to denote any geodesic connecting p and q. Let $\delta \geq 0$; a geodesic metric space X is called δ-*hyperbolic* if for every geodesic triangle $\Delta = \Delta(p, q, r)$ any one side $[p, q]$ of Δ is contained in the δ-neighborhood $\mathcal{N}_\delta([p, r] \bigcup [q, r])$ of the union of the two other sides $[p, r]$ and $[q, r]$. A geodesic metric space is called *hyperbolic* if it is δ-hyperbolic for some $\delta \geq 0$. Examples of hyperbolic metric spaces include metric trees and n-dimensional hyperbolic space $\mathbb{H}^n$. The Euclidean plane provides a typical example of non-hyperbolic metric space. An *isometry* of a metric space (X, d_X) is a self-homeomorphism ϕ of X preserving the distance d_X. If X is a hyperbolic space, then there are three types of isometries of X: elliptic, parabolic, and hyperbolic. For our purpose, we only consider here hyperbolic ones. In the following definition, the ring of integers $\mathbb{Z}$ is equipped with the induced distance from the ordinary distance on $\mathbb{R}$, and we recall that a map $\phi \colon (X, d_X) \to (Y, d_Y)$ between two metric spaces is called a *quasi-isometric embedding* if for some $K \geq 1$ and $L \geq 0$,

$$\frac{1}{K} d_X(x, y) - L \leq d_Y(\phi(x), \phi(y)) \leq K d_X(x, y) + L, \text{ for all } x, y \in X.$$

It is called a *quasi-isometry* if moreover there exists $D \geq 0$ such that for every $y \in Y$ there exists $x \in X$ such that $d_Y(\phi(x), y) \leq D$.

Definition 8.5 (See Gromov [1987].) Let (X, d_X) be a hyperbolic metric space and x_0 be a base point in X. An isometry γ of (X, d_X) is called *hyperbolic* if the orbit map $\mathbb{Z} \to X \colon n \mapsto \gamma^n \cdot x_0$ is a quasi-isometric embedding.

Hyperbolic isometries are generalizations of hyperbolic isometries in classical hyperbolic geometry. In hyperbolic geometry, every hyperbolic isometry preserves a geodesic line. This is no longer true for general δ-hyperbolic metric spaces. Nevertheless, according to the above definition, every hyperbolic isometry of a δ-hyperbolic metric space preserves a quasi-geodesic line which will be called a *quasi-axis*.

8.2.2 Hyperbolic Groups

Let G be a finitely generated group with a finite generating set S. We assume that S is a symmetric set, that is, $S = S^{-1}$. The *Cayley graph* $Cay(G, S)$ of G with respect to S is the metric graph defined as follows. The set of vertices is G and two vertices $g \in G$ and $h \in G$ are connected by an edge if there is an element $s \in S$ such that $g = hs$. The metric is the length metric induced by assigning length one to each edge. Hence $Cay(G, S)$ is a regular locally finite metric graph and G acts by isometries on $Cay(G, S)$ cocompactly and properly discontinuously.

Definition 8.6 (See Gromov [1987].) A finitely generated group G with respect to a finite generating set S is said to be *hyperbolic* if its Cayley graph $Cay(G, S)$ is a δ-hyperbolic metric space for some $\delta \geq 0$.

Remark 8.7 In the above definition, the Cayley graph $Cay(G, S)$ depends on the choice of the generating set S. However, as Cayley graphs defined by different generating sets are quasi-isometric and hyperbolicity is preserved under quasi-isometry, the hyperbolicity of a group is independent of the choice of a generating set.

Theorem 8.8 (See Alonso et al. [1991].) *Let G be a hyperbolic group. If $g \in G$ is an element of infinite order, then g, as an isometry of $Cay(G, S)$, is hyperbolic.*

8.2.3 Mapping Class Groups

Let $\Sigma = \Sigma_{g,n}$ be a genus g closed, orientable, connected surface with n punctures. The *mapping class group* $\mathrm{MCG}(\Sigma)$ is the group of isotopy classes of orientation-preserving homeomorphisms of Σ. Mapping class groups are finitely presented. See Farb and Margalit [2012] for more on mapping class groups. Mapping class groups, except very few cases, are not hyperbolic groups, but there are hyperbolic spaces encoding their hyperbolic behavior. We are going to introduce them now. We will assume that $3g + n \geq 5$. For $3g + n \leq 4$, either the mapping class group is almost trivial or can be dealt with in a slightly modified way. The *curve graph* $\mathcal{C}(\Sigma)$ of Σ, defined in Harvey [1981], is the graph whose vertices are isotopy classes of essential simple closed curves on Σ (that is, embedded circles in Σ not bounding a disk or a disk with a puncture) and edges are defined by disjointness, namely two vertices a and b are connected by an edge if one can choose simple closed curves α and β in a and b, respectively, such that $\alpha \cap \beta = \emptyset$. We then regard $\mathcal{C}(\Sigma)$ as a metric graph as well by assigning length one to each edge. One can easily check that MCG(Σ) acts on $\mathcal{C}(\Sigma)$ by isometries. We remark here that unlike the Cayley graph of a (finitely generated) hyperbolic group, $\mathcal{C}(\Sigma)$ is not locally finite. We have the following remarkable theorem concerning the negatively curved feature of MCG(Σ). Recall that an element g of MCG(Σ) is called *pseudo-Anosov* if no non-trivial power of g fixes the isotopy class of an essential simple closed curve.

Theorem 8.9 [Masur and Minsky, 1999, theorem 1.1] *For the surface* $\Sigma = \Sigma_{g,n}$ *with* $3g + n \geq 5$, *the curve graph* $\mathcal{C}(\Sigma)$ *is a hyperbolic metric space of infinite diameter. An element* g *of* MCG(Σ) *is hyperbolic if and only if* g *is pseudo-Anosov.*

8.3 WPD and Quasimorphisms

In this section, we will heuristically discuss constructions of quasimorphisms on hyperbolic groups and mapping class groups following Bestvina and Fujiwara [2002] (although the construction on hyperbolic groups was originally carried out in Epstein and Fujiwara [1997]). The reader is encouraged to read that paper for historical notes, references, and more details therein. In this section, we will assume hyperbolic groups to be torsion free.

The starting point of the whole story is Brooks quasimorphisms on free groups; see Example 7.7. We reformulate that construction in the context of group actions. Let $F_n (n \geq 2)$ be a free group of rank n with a free generating set $S = \{a_1^{\pm 1}, \ldots, a_n^{\pm 1}\}$. The Cayley graph $X = Cay(F_n, S)$ is then a tree. Any reduced word $\omega = a_{i_1}^{\pm 1} \cdots a_{i_k}^{\pm 1}$ in F_n corresponds to a geodesic segment in X, starting at the vertex $e \in X$ and ending at the vertex $\omega \in X$. Fix a reduced word ω, hence a geodesic segment in X, and then consider all copies of ω in X, that is, the set $\{g.\omega \subset X : g \in F_n\}$. As X is a tree, every element g in F_n also corresponds to a geodesic σ_g in X, namely, the unique geodesic line connecting e and g. We now define a function f_ω on F_n as follows: $f_\omega(g)$ is the maximal number of non-overlapping copies of ω contained in σ_g. The map $f_\omega(g)$ is not a quasimorphism, but one can slightly modify this function to be a quasimorphism. For this, define $h_\omega(g) = f_\omega(g) - f_{\omega^{-1}}(g)$; then $h_\omega \colon F_n \to \mathbb{Z}$ is indeed a quasimorphism.

Assume that G is a hyperbolic group. We want to play the same game for G and $X = Cay(G, S)$. Fix a simple oriented path $\omega \in X$, that is, an oriented geodesic in X. Although one can also consider copies of ω, we face some difficulties here when we want to define f_ω since there are, in general, many geodesics connecting e to $g \in G$. Nevertheless, this difficulty can be overcome by considering the following modified formula for f_ω:

$$f_\omega^W(g) = |g| - \inf_\beta(|\beta| - W|\beta|_\omega),$$

where $|g|$ is the distance between $e \in X$ and $g = g.e \in X$, W is any positive number less than $|\omega|$, β runs over all quasi-geodesics connecting e and g, and $|\beta|_\omega$ is the maximal number of non-overlapping oriented copies of ω appearing in β. By choosing W appropriately, one can show that $h_\omega^W = f_\omega^W - f_{\omega^{-1}}^W$ is a quasimorphism. We fix such a W and denote h_ω^W simply by h_ω. Notice that, up to now, the entire discussion can be generalized to a group acting on a hyperbolic metric graph.

Now there is a serious issue, namely, unlike the case of free groups, h_ω might be a trivial quasimorphism. And even if h_ω is a non-trivial quasimorphism, h_ω might be a genuine homomorphism, which is trivial in the kernel of the second comparison map. The reader could try h_ω for $G = \mathbb{Z}$ as an exercise.

We now go back to Brooks quasimorphisms on the rank 2 free group $F_2 = F\{a, b\}$. Choose $\omega = ab$, then $f_\omega((ab)^n) = n$ and $f_{\omega^{-1}}((ab)^n) = 0$; this shows that h_ω is non-trivial. Since $h_\omega(a) = h_\omega(b) = 0, h_\omega(ab) = 1$, h_ω is not a homomorphism. One can also check that it is not a homomorphism up to a bounded function. The above discussion in fact contains all we need to define a quasimorphism for a group acting on a hyperbolic metric graph.

Let G be a group acting on a hyperbolic metric graph X by isometries. The following properties are used in the construction of Brooks quasimorphisms:

(1) G does not have $\mathbb{Z}$ as a subgroup of finite index. Namely, G is non-elementary.

(2) G contains a hyperbolic isometry g. Some long piecewise geodesic segment ℓ in an oriented axis σ_g of g (here σ_g is a quasi-geodesic invariant under g) will play the role of $\omega = ab$. In order to show that h_ℓ is non-trivial, g is required to satisfy the condition that one cannot find an arbitrarily large number of copies of ℓ in a bounded neighbourhood of an oriented axis of g^{-1}. Note that as X is a δ-hyperbolic metric space, those bounds only depend on δ. One can also argue as in the construction of Brooks quasimorphisms that h_ℓ is not a homomorphism up to a bounded function.

Bestvina–Fujiwara proposed a condition on the action $G \curvearrowright X$, called *weak proper discontinuity* (*WPD*, for short), which provides infinitely many required hyperbolic isometries in G.

Definition 8.10 [Bestvina and Fujiwara, 2002] Let G be a group acting on a hyperbolic metric graph (X, d_X). The action is said to be *WPD* if the following are all satisfied:

(1) G does not have $\mathbb{Z}$ as a subgroup of finite index.
(2) G contains an element that acts on X as a hyperbolic isometry.
(3) For every element $g \in G$ acting as a hyperbolic isometry on X, every x in X and every $R > 0$, there are positive numbers N and T, depending on g, x, r and such that

$$\left| \left\{ \gamma \in G : d_X(x, \gamma \cdot x) \leq R, d_X(g^N \cdot x, (\gamma g^N) \cdot x) \leq R \right\} \right| \leq T.$$

Theorem 8.11 [Bestvina and Fujiwara, 2002, proposition 6] *Assume that the action of G on X is WPD. Then there exist two hyperbolic elements g_1, $g_2 \in G$ such that there is an oriented long segment ℓ in an oriented quasi-axis of g_1 such that every copy of ℓ is outside a bounded oriented neighbourhood of an oriented quasi-axis of g_2, where the orientation on the neighborhood is induced from the orientation of the quasi-axis.*

The above theorem (together with Bestvina and Fujiwara [2002, proposition 2]) tells us that WPD actions of a group essentially enable us to find two highly independent (with respect to the group action) elements in the group and, using these two elements, one could construct a group element like the element used in the construction of Brooks quasimorphisms on free groups. Interestingly, these two group elements actually promise the existence of infinitely many linearly independent quasimorphisms on the group, which shows the kernel of the second comparison map even to be uncoutably dimensional.

We conclude our discussion with two examples. One could refer to Bestvina and Fujiwara [2009] for more applications of WPD.

Example 8.12 (See Epstein and Fujiwara [1997].) Let G be a non-elementary, torsion-free hyperbolic group. As $Cay(G, S)$ is uniformly locally finite and every non-trivial element in G is hyperbolic by Theorem 8.8, it is easy to verify that the action G on its Cayley graph $Cay(G, S)$ satisfies WPD. Hence, the kernel of the second comparison map c^2 is of infinite dimension.

Example 8.13 (See Bestvina and Fujiwara [2002].) Let $\Sigma = \Sigma_{g,n}$ be a compact, orientable, connected surface of genus g with n punctures such that $3g + n \geq 5$, and let $\mathcal{C}(\Sigma)$ be the curve graph of Σ. Then by Theorem 8.9, the curve graph $\mathcal{C}(\Sigma)$ is hyperbolic and pseudo-Anosov mapping classes act as hyperbolic isometries. Proposition 11 in Bestvina and Fujiwara [2002] shows that the action of $\mathrm{MCG}(\Sigma)$ on $\mathcal{C}(\Sigma)$ also satisfies WPD. Thus the kernel of the second comparison maps c^2 for $\mathrm{MCG}(\Sigma)$ is of infinite dimension as well. In fact, this result holds also for any subgroup of $\mathrm{MCG}(\Sigma)$ which is not virtually abelian.

Acknowledgments

The author would like to thank Indira Chatterji for correcting the manuscript, and Benjamin Zarka and Lamine Messaci for their interest and comments.

9

Extension of Quasicocycles from Hyperbolically Embedded Subgroups

Francesco Fournier-Facio

We have seen in Chapter 8 that one can construct non-trivial quasimorphisms on groups acting nicely on hyperbolic spaces by generalizing the Brooks construction in free groups. Here we will see another way in which one can exploit the knowledge of the bounded cohomology of the free group to prove non-vanishing results for much more general groups, even in higher degrees. Namely, we will explain how to extend a quasicocycle (these are higher-dimensional analogs of quasimorphisms) from a subgroup to the ambient group, under the condition that the subgroup is *hyperbolically embedded*, a notion introduced by Dahmani, Guirardel, and Osin [Dahmani et al., 2017]. It turns out that a very large class of groups, called *acylindrically hyperbolic*, has hyperbolically embedded subgroups of the form $F \times K$, where F is a free group and K is a finite group. It follows that the bounded cohomology of such groups is "at least as large" as that of the free group, in all degrees.

This construction was carried out in degree 2 by Hull and Osin [2013], and then by Frigerio, Pozzetti, and Sisto in higher degrees [Frigerio et al., 2015]. Here we will follow the latter approach.

9.1 Alternating Quasicocycles

The definition of quasimorphisms we saw in the previous two chapters uses the definition of bounded cohomology in terms of the *bar resolution*. For our purposes, it will be more useful to work with the *bounded homogeneous resolution* (Definition 11), which we denote by $(C_b^\bullet(G)^G, \delta^\bullet)$. We further denote *bounded cocycles* by $Z_b^n(G) = \ker(\delta^n)$ and *bounded coboundaries* by $B_b^n(G) = \operatorname{im}(\delta^{n-1})$, so that the *bounded cohomology of G* with trivial real coefficients is $H_b^n(G) = Z_b^n(G)/B_b^n(G)$.

An element of G^{n+1} is called a *simplex*. If $S \subset G$ is any set, an element of $S^{n+1} \subset G^{n+1}$ is said to be *supported on S*. We can extend cochains by linearity on formal sums of simplices: these are called *chains* and the set of all chains is denoted by $C_n(G)$. Similarly, a chain is supported on S if all of the simplices appearing in its expression are supported on S; their collection is denoted by $C_n(S)$. There is a natural *boundary* operator on $C_n(G)$ that is the pre-dual of the coboundary operator δ:

$$\partial_{n+1}(g_0, \ldots, g_{n+1}) := \sum_{i=0}^{n+1} (-1)^i (g_0, \ldots, \hat{g}_i, \ldots, g_{n+1}).$$

By *predual* we mean that $\varphi \circ \partial_{n+1} = \delta^n(\varphi)$ for any n-cochain φ. Note that the boundary operator preserves the support of a chain, namely $\partial_{n+1}(C_{n+1}(S)) \subset C_n(S)$.

Dropping all of the boundedness conditions, we have the complex of invariant cochains $(C^n(G)^G, \delta^\bullet)$, which defines the *(ordinary) cohomology of G* with trivial real coefficients $H^n(G)$.

For our purposes, it will be convenient to disregard the ordering of the entries $(g_0, \ldots, g_n)$ but only consider the set $\{g_0, \ldots, g_n\}$ (up to sign). Moreover, it will be useful to ignore *degenerate simplices*, namely those in which some vertex appears more than once, which we see as "lower-dimensional". For this, we will consider *alternating cochains* (see Remark 13), that is cochains φ satisfying

$$\varphi(g_{\sigma(0)}, \ldots, g_{\sigma(n)}) = \text{sign}(\sigma)\varphi(g_0, \ldots, g_n)$$

for any permutation $\sigma \in S_{n+1} \cong Sym\{0, \ldots, n\}$. Note that an alternating cochain vanishes on degenerate simplices. Restricting to alternating cochains defines the subcomplexes $(C_{alt}^\bullet(G)^G, \delta^\bullet)$ and $(C_{b,alt}^\bullet(G)^G, \delta^\bullet)$, which still compute the cohomology and bounded cohomology of G [Frigerio, 2017, 4.10]. This is because there is a G-equivariant projection $C_b^\bullet(G) \to C_{b,alt}^\bullet(G)$, which is defined by averaging: given a simplex $\overline{g} = (g_0, \ldots, g_n) \in G^{n+1}$, we define

$$\text{alt}_n(\overline{g}) = \frac{1}{(n+1)!} \sum_{\sigma \in S_{n+1}} \text{sign}(\sigma)(g_{\sigma(0)}, \ldots, g_{\sigma(n)}).$$

This extends to an operator $\text{alt}_n \colon C_n(G) \to C_n(G)$, whose dual $\varphi \mapsto \varphi \circ \text{alt}_n$ defines a projection $C^n(G) \to C_{alt}^n(G)$, which preserves everything we want: cocycles, coboundaries, boundedness, invariance, quasicocycles, and so on. Also notice that $\text{alt}_n(\overline{g}) = 0$ if and only if $\overline{g}$ is a degenerate simplex. Analogously, we call an n-chain *degenerate* if it is in the kernel of alt_n.

Recall that our goal is to extend bounded cohomology classes from $H_b^n(H)$ to $H_b^n(G)$, where H is a subgroup of G. We will try and do this at the level of cochains: in the best case scenario, we would have a chain map $\Theta : C^n_{alt}(H)^H \to C^n_{alt}(G)^G$ which sends bounded cochains to bounded cochains, such that restricting $\Theta(\varphi)$ to H^{n+1} gives back φ. Such a map would allow us to extend any bounded cohomology class in H to one in G. But we are working in geometric group theory, and so we do not want to ask for something so precise, which can only work in extremely restrictive settings. So the more reasonable thing to ask for is that Θ is an *approximate chain map* (that is, $\delta^n \Theta^n$ is at a bounded distance from $\Theta^{n+1}\delta^n$), and restricting $\Theta(\varphi)$ to H^{n+1} gives a cochain at a bounded distance from φ.

The biproduct of this "quasification" is that now the image of a cocycle under Θ is not necessarily a cocycle. So we will also "quasify" the notion of cocycle. An *n-quasicocycle* is an n-cochain φ whose coboundary $\delta^n(\varphi)$ is bounded. We denote by $QZ^n_{alt}(G)^G$ the space of alternating invariant quasicocycles. Therefore if $\varphi \in QZ^n_{alt}(G)$, then $\delta^n(\varphi)$, being a bounded cocycle, defines a bounded cohomology class in $H^{n+1}_b(G)$; we call such classes *exact* and denote the corresponding subspace $EH^{n+1}_b(G)$. Note that $\delta^n(\varphi)$ is always a coboundary, but it is not necessarily a coboundary of a bounded cochain: this only happens if φ is a *trivial* quasicocycle, that is, it is at a bounded distance from a true cocycle. Now the conditions on Θ ensure that it not only sends a cocycle to a quasicocycle, but it even sends a quasicocycle to a quasicocycle. Moreover, a non-trivial quasicocycle will be sent to a non-trivial one, although a priori even a trivial one can be sent to a non-trivial one. Therefore we can extend any non-zero *exact* bounded cohomology class in H to one in G. If H is, say, a non-abelian free group, then any class in degree of at least 2 is exact (because the ordinary cohomology vanishes), and so knowing that its bounded cohomology is large implies that the exact bounded cohomology of G is large.

Let us connect this notion of quasicocycles with the more familiar notion of quasimorphisms we saw in Chapters 7 and 8. Given an invariant 1-quasicocycle $\varphi \in QZ^1(G)^G$, define $\tilde{\varphi} \colon G \to \mathbb{R}, g \mapsto \varphi(1, g)$. We claim that this is a quasimorphism. Indeed:

$$|\tilde{\varphi}(gh) - \tilde{\varphi}(g) - \tilde{\varphi}(h)| = |\varphi(1, gh) - \varphi(1, g) - \varphi(1, h)| =$$

$$= |\varphi(1, gh) - \varphi(1, g) - \varphi(g, gh)| = |\delta^1(\varphi)(1, g, gh)| \leq \|\delta^1(\varphi)\|_\infty,$$

where in the second equality we used φ is G-invariant. Therefore $D(\tilde{\varphi}) \leq \|\delta^1(\varphi)\|_\infty$ and $\tilde{\varphi}$ is a quasimorphism.

Conversely, given a quasimorphism $\tilde{\varphi}$, define $\varphi \colon G^2 \to \mathbb{R}; (g, h) \mapsto \tilde{\varphi}(g^{-1}h)$. Then the same calculation shows that φ is an invariant 1-quasicocycle.

9.2 Intuitive Idea

We start to construct the map Θ, so as to identify what conditions we need for the groups $H \leq G$ in order to make things work. Fix $\varphi \in C_{alt}^n(H)^H$. Given a simplex $\overline{g} = (g_0, \ldots, g_n) \in G^{n+1}$, we want to define $\Theta(\varphi)(\overline{g})$ in terms of φ. But φ only knows what to do with values in H, so we should assign to $\overline{g}$ some simplex supported on H, in a meaningful way. We could try and put a suitable word metric on G, and we should be able to project nicely the vertices of $\overline{g}$ onto H under some negative curvature assumption (which is typically what makes projections well-behaved). It is a bit too restrictive to require that these closest-point projections be single-valued, but that is not a big deal since we know how to evaluate φ at chains supported on H, not just simplices, so averaging over all possible projections – as long as there are finitely many – is something we can do. However, we have another problem: if we only project onto H, it is unlikely that the resulting $\Theta(\varphi)$ would be G-equivariant.

Still, we can do something which is slightly more complex but does preserve equivariance: instead of looking only at H, we look at all *cosets* $B \in G/H$. There is an obvious way to define φ on a simplex supported on a coset, by treating it as if it were a equivariant function (which it is, on H!). Namely if $(b_0, \ldots, b_n) \in B^{n+1}$, we can define

$$\varphi_B(b_0, \ldots, b_n) := \varphi(1, b_0^{-1}b_1, \ldots, b_0^{-1}b_n).$$

This is well-defined: since b_0 and b_i are in the same coset, $b_0^{-1}b_i \in H$, and so it makes sense to evaluate φ on this tuple. Moreover, it is easy to check that this is still alternating, and that $\varphi_H = \varphi$. Then φ_B can be evaluated on any chain supported on B.

Now that we know how to work with chains supported on cosets of H, the more reasonable task is to define $\Theta(\varphi)$ by assigning to $\overline{g}$ some chain supported on a coset B, for any $B \in G/H$. So what we need are **trace operators** $\text{tr}_n^B : C_n(G) \to C_n(B)$ defined on simplices by averaging over closest-point projections, and then we can define

$$\Phi(\overline{g}) := \sum_{B \in G/H} \varphi_B(\text{tr}_n^B(\overline{g})).$$

This may well be a reasonable thing to ask if H had finite index in G, that is, if the sum above had only finitely many terms. But recall our ultimate goal: we want to extend quasicocycles from H to G, and in our applications H will be a virtually free group. So if we restricted to this setting, then our results would only apply to virtually free groups, which would be a little disappointing. In the general case, the sum above is infinite and so it has no guarantee of converging. Moreover, we are working with things up to bounded distance, so it would be awkward if we found ourselves dealing with convergence.

To solve these issues, we should be able to identify cosets that are **relevant** for $\overline{g}$: those which together capture enough about $\overline{g}$ to be able to forget the rest while paying a small price. In whatever way we define them, the set $\mathcal{R}(\overline{g})$ of relevant cosets for $\overline{g}$ should be finite, and we should define our trace operators to evaluate to 0 on non-relevant cosets: that is, $\mathrm{tr}^n_B(\overline{g}) = 0$ for all $B \notin \mathcal{R}(\overline{g})$. Then the sum above becomes finite, and we have a well-defined map $\Phi \colon G^{n+1} \to \mathbb{R}$.

Moreover, it makes sense that if $\overline{g}$ is already supported on B, then the only coset that can be relevant for $\overline{g}$ should be B itself, and that in that case $\mathrm{tr}^B_n(\overline{g})$ should be just $\overline{g}$. Our notion of relevant cosets will be defined in terms of a bounded error, and while it is true that $\mathcal{R}(B^{n+1}) \subset \{B\}$, there will be some simplices supported on B for which there are no relevant simplices at all. These are simplices that are "too small" to be visible under this coarse approach, so we will inventively call them **small**, and analogously a chain in $C_n(B)$ will be **small** if the simplices appearing in it are small. Now if $\overline{g} \in H^{n+1}$ is not small, then $\Phi(\overline{g}) = \varphi(\overline{g})$, and so Φ is indeed an extension of φ, up to small simplices. For the moment, we write $\Theta^n(\varphi) := \Phi$ as a good candidate for the extension, and keep in mind that this works "up to small chains."

Until now, we have used no additional property of φ. It is easy to believe that Φ will be G-invariant if φ is H-invariant. The next thing we would like the extension to preserve is *boundedness*. So suppose that φ is bounded, and so φ_B is also bounded for all B, with the same bound. We know that the sum defining Φ is finite, but in order to have Φ be bounded itself, this is not enough: we need a uniform bound on the number of non-zero terms in the sum. We will take advantage of another fact that we have not used yet, namely that φ is alternating, and so φ_B is alternating too. Intuitively, what tr^B_n is doing is projecting simplices in G onto simplices in B, so geometrically we should expect there to be many cosets for which $\mathrm{tr}^B_n(\overline{g})$ is "lower-dimensional," that is, *degenerate*. But whenever $\mathrm{tr}^n_B(\overline{g})$ is degenerate, we have $\varphi_B(\mathrm{tr}^B_n(\overline{g})) = 0$. So what we will have to show is that there is a uniform bound $C(n)$ such that for any simplex $\overline{g} \in G^{n+1}$ there are at most $C(n)$ cosets B for which $\mathrm{tr}^B_n(\overline{g})$ is non-degenerate.

The next and most important thing that we would like the extension to preserve is quasicocycles. As we mentioned in the beginning, this will be true if Θ commutes with the coboundary operator up to bounded errors. Since we defined Θ as some sort of dual of the trace operator, this is the same as asking that tr_n^B commutes with the boundary operator up to bounded errors. The problems are caused by the difference

$$\partial_n \mathrm{tr}_n^B(\overline{g}) - \mathrm{tr}_{n-1}^B \partial_n(\overline{g}) \in C_n(B),$$

as in: if this difference were zero, then we would be done. But remember that we have already had to introduce a notion of *small chain*, and it turns out that this difference will be small according to our definition.

Now it seems that everything works "up to small chains," at the level of chains. But how do we make everything work up to bounded errors, at the level of cochains? One thing we can do is to simply ignore small chains in our definition of Φ. That is, we define φ_B' to vanish on small simplices, and coincide with φ_B everywhere else, and then define Φ in terms of φ_B' instead. This solves the problem of small chains, but of course we do not want to change things too much. More precisely, we would like that the quantity

$$K(\varphi) := \sup\{|\varphi(\overline{h})| : \overline{h} \in H^{n+1} \text{ is small}\},$$

which controls the difference between φ_B and φ_B', is finite, even if φ is not bounded. Whatever definition of small we end up choosing, it makes sense that small simplices supported on H should be H-invariant. Since φ is H-invariant, $K(\varphi)$ may also be computed by only looking at simplices containing the identity. So what we really want is that there should be only finitely many small simplices containing the identity: some sort of *local finiteness* condition.

To sum up, we have the following checklist:

Checklist 9.1 Make the following notions more precise:

1. A notion of *relevant coset* for a simplex $\overline{g} \in G^{n+1}$, such that
   1. The set $\mathcal{R}(\overline{g})$ of cosets relevant for $\overline{g}$ is finite;
   2. If $\overline{g}$ is supported on B, then $\mathcal{R}(\overline{g}) \subset \{B\}$.
2. A notion of *small simplex* and the corresponding notion of *small chain* supported on a given coset, such that
   1. Small simplices supported on H are H-invariant;
   2. There are only finitely many small simplices supported on H and containing the identity.
3. A *trace operator* $\mathrm{tr}_n^B : C_n(G) \to C_n(B)$ for any $B \in G/H$, such that
   1. $\mathrm{tr}_n^B(\overline{g}) = 0$ if B is *not* relevant for $\overline{g}$;

2. $\mathrm{tr}_n^B(\overline{g}) = \overline{g}$ for all $\overline{g} \in B^{n+1}$, unless $\overline{g}$ is small;
3. There exists a uniform bound $C(n)$, such that for a simplex $\overline{g}$ there are at most $C(n)$ cosets B such that $\mathrm{tr}_n^B(\overline{g})$ is non-degenerate;
4. $(\partial_n \mathrm{tr}_n^B - \mathrm{tr}_{n-1}^B \partial_n)(\overline{g}) \in C_n(B)$ is a small chain, for all $\overline{g} \in G^{n+1}$.

9.3 Hyperbolically Embedded Subgroups

To define the trace operator, we want to have some way to project the vertices of a simplex onto a coset. Closest-point projections typically behave well under some negative curvature assumptions, for instance in hyperbolic groups. As a matter of fact, we do not need hyperbolicity itself, but only for H to be embedded into G in a way that retains some kind of hyperbolicity. This is done by the notion of a *hyperbolically embedded subgroup*, as introduced by Dahmani, Guirdardel, and Osin [Dahmani et al., 2017]. In this section, we define it and look at some properties of closest-point projections that lead very naturally to the notions of relevant cosets and small simplices that we need to define.

9.3.1 Definition and Examples

Let us fix for the rest of this section a group G and a subgroup H, and denote by $H^* := H \setminus \{1\}$. A (possibly infinite) subset $X \subset G$ is a *relative generating set* if G is generated by $X \sqcup H^*$. We denote by $\mathrm{Cay}(G, X \sqcup H^*)$ the corresponding Cayley graph: note that we are taking a disjoint union, so if $x \in X$ is equal to $h \in H^*$, there are going to be two distint edges from g to $gx = gh$: one labeled by x and one by h. Here H itself has diameter 1, since every element in H is an edge. This is not very interesting, so when looking at H we will consider a *relative metric*. For $g, h \in H$, this is denoted by $d_H(g, h)$ and is the shortest length of a path from g to h in $\mathrm{Cay}(G, X \sqcup H^*)$, which does not use any "shortcut," that is, any edge labeled by an element of H and connecting two elements of H. Similarly, for any coset $B \in G/H$, we can translate the relative metric d_H to a metric d_B on B.

Definition 9.2 [Dahmani et al., 2017, definition 2.1] We say that H is *hyperbolically embedded in* (G, X) if $\mathrm{Cay}(G, X \sqcup H^*)$ is hyperbolic and the metric space (H, d_H) is locally finite (i.e., balls of finite radius are finite). We say that H is *hyperbolically embedded in* G if it is hyperbolically embedded in (G, X) for some relative generating set X. We write $H \hookrightarrow_h (G, X)$ and $H \hookrightarrow_h G$.

Since we are not asking for X to be finite, the second condition is there to restrict this notion from being too general. Here is a non-example that exhibits the importance of the local finiteness condition:

Example 9.3 Let H be any group, $G = H \times \mathbb{Z}$ and $X = \{x\}$ a generator of $\mathbb{Z}$. Then $\mathrm{Cay}(G, H^* \sqcup X)$ is quasi-isometric to a line and thus hyperbolic. But given $h_1, h_2 \in H$ we have a "shortcut-free" path of length 3 connecting them, namely $h_1 \to xh_1 \to xh_2 \to h_2$. So (H, d_H) can only be locally finite if H is already finite.

Here are the two simplest examples:

Example 9.4 1. $G \hookrightarrow_h (G, \emptyset)$. Indeed, the corresponding Cayley graph has diameter 1, so it is hyperbolic, and there is no way to connect elements of G without using edges labeled by elements of G, so d_G-balls of finite radius are just singletons.

2. Let H be a finite subgroup of G; then $H \hookrightarrow_h (G, G)$. Indeed, the corresponding Cayley graph has diameter 1, so it is hyperbolic, and H is finite so any metric on it will be locally finite.

These examples should be considered as trivial, in the same way that a geodesic metric space of diameter one is the trivial example of a hyperbolic space. If $H \hookrightarrow_h G$ is of this form, we say that H is *degenerate*.

This definition looks a lot like that of relatively hyperbolic groups, but the key difference is that here the relative generating set is allowed to be infinite:

Proposition 9.5 [Dahmani et al., 2017, proposition 2.4] *G is hyperbolic relative to H if and only if $H \hookrightarrow_h (G, X)$ for a finite relative generating set $X \subset G$.*

Allowing X to be infinite gives this notion a lot more flexibility. We already saw this in the second degenerate case: any finite subgroup is hyperbolically embedded, but a group is hyperbolic relative to a finite subgroup if and only if it is already hyperbolic.

Another interesting example of this is in mapping class groups: these are almost never relatively hyperbolic, but they admit non-degenerate hyperbolically embedded subgroups (except for a finite number of exceptional surfaces). This is part of a more general fact that relates nicely to Chapter 8: recall the definition of a weakly properly discontinuous (WPD) element (Definition 8.10). Then we have:

Theorem 9.6 [Dahmani et al., 2017, corollary 2.9] *Let G be a group acting on a hyperbolic space, and let g ∈ G be loxodromic and WPD. Then g is contained in a unique maximal virtually cyclic subgroup $E(g)$, and $E(g) \hookrightarrow_h G$.*

Recall from Chapter 8 (Theorem 8.9) that given a (possibly punctured) orientable closed surface Σ, its mapping class group acts on the curve graph, which is hyperbolic, and pseudo-Anosov elements are loxodromic and WPD. So we obtain:

Theorem 9.7 [Dahmani et al., 2017, corollary 2.19] *Let $g \in \mathrm{MCG}(\Sigma)$ be a pseudo-Anosov element. Then g is contained in a unique maximal virtually cyclic subgroup $E(g)$ and $E(g) \hookrightarrow_h G$.*

To grasp just how general this definition is, let us mention two more examples of the same flavor, but in different contexts.

Theorem 9.8 [Dahmani et al., 2017, corollary 2.20; Sisto, 2018, theorem 1.1 and theorem 1.3] *Let G be a group and $g \in G$. In the following cases, g is contained in a unique maximal virtually cyclic subgroup $E(g)$, and $E(g) \hookrightarrow_h G$:*

1. *(Dahmani–Guirardel–Osin) F is a non-abelian free group, $G = Out(F)$, and $g \in Out(F)$ is irreducible with irreducible powers.*
2. *(Sisto) G is a group acting properly on a proper $CAT(0)$ space, and $g \in G$ is a rank one isometry.*

It is not important to know the precise definitions involved in this theorem; what matters is that there are lots of interesting groups with non-degenerate hyperbolically embedded subgroups.

At this point, one may be noticing a disappointing pattern – namely all of our examples of non-degenerate hyperbolically embedded subgroups are virtually cyclic. These are amenable, so their bounded cohomology is trivial in positive degrees. One could use virtually cyclic subgroups to obtain classes in degree 2 [Hull and Osin, 2013], but it would require some more work to show that these are non-trivial, and moreover we would like to tackle higher degrees as well. So it would be much better to have hyperbolically embedded subgroups with large bounded cohomology. This is taken care of by

Theorem 9.9 [Dahmani et al., 2017, theorem 2.24] *Let G be a group containing a non-degenerate hyperbolically embedded subgroup. Then G contains a unique maximal finite normal subgroup K, and there exist subgroups $H = F \times K \le G$, where F is a free group of arbitrary rank, such that $H \hookrightarrow_h G$.*

So a group admits a non-degenerate hyperbolically embedded subgroup if and only if it admits one of the form $F \times K$, with F non-abelian free and K finite. These have large bounded cohomology, so they suit our purposes.

Let us at least mention that there is another notion that is (very) often used instead of this one, namely that of *acylindrical hyperbolicity*. We will not define it here but refer the reader to Osin [2016]. Suffice it to say that a group is acylindrically hyperbolic if and only if it admits a non-degenerate hyperbolically embedded subgroup, and in fact there is yet another equivalent condition in terms of the WPD property from Chapter 8 [Osin, 2016, theorem 1.2].

9.3.2 Projections

Now that we saw that the notion of hyperbolically embedded subgroups is so general, it is time to check that it is actually useful for our purposes. Recall that we want to define a trace operator that assigns to a simplex in G a chain supported on a coset B, and that we hope to do it using *closest-point projections*. The Cayley graph $\mathrm{Cay}(G, X \sqcup H^*)$ can be seen as the (possibly disconnected) Cayley graph $\mathrm{Cay}(G, X)$ to which we add a shortcut between any two elements in the same coset B, given by an edge of length one labeled by an element of H. We will actually work on a different metric graph [Frigerio et al., 2015, definition 2.3], denoted $(\hat{G}, \hat{d})$ and obtained starting from $\mathrm{Cay}(G, X)$, adding a vertex $c(B)$ for every coset $B \in G/H$, and an edge of length 1/4 from any $g \in B$ to $c(B)$. Now the shortcuts between elements of B have length 1/2, and they pass through the cone vertex $c(B)$. These cone points make projections more "stable," as we will shortly see in Lemma 9.11.

Definition 9.10 Let $B \in G/H$ and x be a vertex of $\hat{G}$. We define the *projection of x onto B* to be the set

$$\pi_B(x) := \{y \in B : \hat{d}(x, y) = \hat{d}(x, B)\}.$$

For a set S we denote $\pi_B(S) := \bigcup_{x \in S} \pi_B(x)$.

Note that there is no reason for $\pi_B(x)$ to be a singleton. For example, $\pi_B(c(B))$ is all of B. If $x \neq c(B)$ is not one of these cone vertices, then $y \in B$ realizes the minimal distance if and only if $x \to y \to c(B)$ is a geodesic path from x to $c(B)$, of which there could be many. On the other hand, if $x \in G$, then $\pi_B(x)$ is finite. To see this, recall that we have a relative metric d_B on any $B \in G/H$; we denote by $diam_B$ the diameter with respect to this metric. We can bound $diam_B(\pi_B(x))$ in terms of $d(x, B)$, and since (B, d_B) is locally finite by the definition of hyperbolically embedded subgroup, this implies that $\pi_B(x)$ is itself finite.

The next lemma is the key property of projections in the context of hyperbolically embedded subgroups that we need in order to turn the notions from Section 9.2 into precise definitions 9.2.

Lemma 9.11 [Frigerio et al., 2015, lemma 2.8] *There exists a constant $D \geq 1$ with the following property. For any $x, y \in \hat{G}$ and any $B \in G/H$, if $diam_B(\pi_B(x) \cup \pi_B(y)) \geq D$, then all geodesics from x to y pass through $c(B)$.*

In other words, if x and y project far apart on B, then the fastest way to connect them is using the shortcut through $c(B)$. The special thing is that this is true of *all* geodesics joining x to y. Whatever definition of *relevant* we will land on, it makes sense now that the coset B is relevant for the 2-simplex (x, y). So this lemma is really taking us in the right direction: in fact, we will define relevant cosets and small simplices in terms of the constant D in the lemma.

We sketch the proof, which exhibits clearly how hyperbolicity enters the picture.

Sketch of proof For simplicity, let us assume that $x, y \in G$ (so they are not of the form $c(B')$), and let $\hat{\gamma}$ be a geodesic connecting them in $(\hat{G}, \hat{d})$. Let $\hat{\gamma}_x$ be a geodesic from x to any $x_B \in \pi_B(x)$; choose similarly $y_B \in \pi_B(y)$ and $\hat{\gamma}_y$. Let e denote the edge in $\text{Cay}(G, X \sqcup H^*)$ labeled by an element of H^* connecting x_B to y_B. We assume that $\hat{\gamma}$ does not pass through $c(B)$ and will prove that $d_B(x_B, y_B)$ is uniformly bounded. Since x_B and y_B were arbitrary, this implies that $diam_B(\pi_B(x) \cup \pi_B(y))$ is uniformly bounded, and we are done.

We modify $\hat{\gamma}$ to obtain a path γ in $\text{Cay}(G, X \sqcup H^*)$ by replacing any shortcut of length $1/2$ going through $c(B')$, for some $B' \in G/H$, by a shortcut of length one connecting two elements of B' and labeled by some element of H^*. We do the same with $\hat{\gamma}_{x,y}$ obtaining $\gamma_{x,y}$. Then one can easily show that this is a quasi-geodesic: all we have done is replace shortcuts of length $1/2$ by shortcuts of length 1.

So now we have a quasi-geodesic quadrangle in the hyperbolic space $Cay(G, X \sqcup H^*)$, with vertices $\{x, x_B, y_B, y\}$ and edges $\{\gamma_x, e, \gamma_y, \gamma\}$. This implies that any point on this quadrangle is C-close to one of the other three edges, for some C depending only on the hyperbolicity constant. This jumping around in short distances between edges allows us to construct a cycle containing e, of length bounded in terms of C, and such that the only edge connecting two elements of B is e itself (recall that we are assuming that $\hat{\gamma}$ does not go through $c(B)$). Taking out e defines a path from x_B to y_B that does not use any shortcut in B, and so its length, which is uniformly bounded, bounds $d_B(x_B, y_B)$. $\square$

9.4 The Trace Operator

In this section we go through Checklist 9.1: we define precisely the notions mentioned in Section 9.2 that make the construction work, and sketch the proof that these verify the conditions therein: all proofs are taken from Frigerio et al. [2015, section 3]. We start with the notion of relevant coset. As we mentioned after Lemma 9.11, if $x, y \in G$, and $B \in G/H$ satisfy the condition in the statement (namely $diam_B(\pi_B(x) \cup \pi_B(y)) \geq D$), then B should be relevant for the simplex (x, y). The definition of relevant coset is slightly stricter but that is the idea:

Definition 9.12 Let $\overline{g} \in G^{n+1}$. Then $B \in G/H$ is *relevant* for $\overline{g}$, denoted $B \in \mathcal{R}(\overline{g})$, if $diam_B(\pi_B(\overline{g})) \geq 2D$.

Let us check that this notion satisfies conditions $1(a)$ and $1(b)$ of Checklist 9.1:

Lemma 9.13 *The set $\mathcal{R}(\overline{g})$ is finite. Moreover, if $\overline{g} \in B^{n+1}$, then $\mathcal{R}(\overline{g}) \subset \{B\}$.*

Proof Note that $\mathcal{R}(\overline{g}) = \cup_{i \neq j} \mathcal{R}(g_i, g_j)$, so it suffices to check finiteness for $(x, y) \in G^2$. If B is relevant for (x, y), then any geodesic from x to y in $(\hat{G}, \hat{d})$ goes through $c(B)$, which implies that the cardinality of $\mathcal{R}(\overline{g})$ is at most the number of vertices in $\hat{\gamma}$. For the second statement, the only geodesic joining $x, y \in B$ in $(\hat{G}, \hat{d})$ is $x \to c(B) \to y$, since it is the only one of length $1/2$. $\square$

Given $\overline{g} \in B^{n+1}$, the proof only shows that $\mathcal{R}(\overline{g}) \subset \{B\}$: the left-hand side could be empty. This leads naturally to the definition of the small simplex:

Definition 9.14 A simplex $\overline{g} \in B^{n+1}$ is *small* if $\mathcal{R}(\overline{g}) = \emptyset$. Equivalently, if $diam_B(\overline{g}) < 2D$.

We now check that this notion satisfies conditions $2(a)$ and $2(b)$ of Checklist 9.1:

Lemma 9.15 *The set of small simplices supported on H is H-invariant, and there are only finitely many such simplices containing the identity.*

Proof If $\overline{g} \in H^{n+1}$ is small and $h \in H$, then $diam_H(h\overline{g}) = diam_H(\overline{g}) < 2D$ and so $h\overline{g}$ is small. The second statement follows directly from the local finiteness of the metric d_H: the ball of radius $2D$ around the identity is finite. $\qquad\qquad\qquad\qquad\qquad\qquad\qquad\qquad\qquad\qquad\qquad\qquad\qquad\qquad$ $\square$

Moving on to Item 3 of Checklist 9.1, we define the trace operator $\mathrm{tr}_n^B : C_n(G) \to C_n(B)$ by doing exactly what we said: let it vanish on non-relevant cosets, and otherwise average over all possible projections (recall that $\pi_B(x)$ is finite for all $x \in G$ and all $B \in G/H$).

Definition 9.16 Let $\overline{g} = (g_0, \dots, g_n) \in G^{n+1}$ and $B \in G/H$. We define $\mathrm{tr}_n^B(\overline{g}) = 0$ if $B \notin \mathcal{R}(\overline{g})$, and otherwise

$$\mathrm{tr}_n^B(\overline{g}) := \frac{1}{\prod_{j=0}^n |\pi_B(g_j)|} \sum_{h_j \in \pi_B(g_j)} (h_0, \dots, h_n) \in C_n(B).$$

Condition $3(a)$ of Checklist 9.1 is in the definition. For condition $3(b)$, if $\overline{g} \in B^{n+1}$, then we saw that $\mathcal{R}(\overline{g}) \subset \{B\}$ and moreover $\pi_B(g_i) = g_i$, clearly. So $\mathrm{tr}_n^B(\overline{g}) = 0$ if $\overline{g}$ is small, and $\mathrm{tr}_n^B(\overline{g}) = \overline{g}$ otherwise. The next thing to check is condition $3(c)$; the proof is, however, more technical than the others, so we refer the reader to Frigerio et al. [2015, proposition 3.9]. Suffice it to know that here it becomes clear why we require the definition of relevant cosets to be with the stricter "$\geq 2D$" instead of just "$\geq D$".

We are left to prove condition $3(d)$ from Checklist 9.1:

Lemma 9.17 *Let $n \geq 2$ and $\overline{g} \in G^{n+1}$. Then $(\partial_n \mathrm{tr}_n^B(\overline{g}) - \mathrm{tr}_{n-1}^B \partial_n(\overline{g})) \in C_{n-1}(B)$ is a small chain for any $B \in G/H$.*

Sketch of proof Let $(*)$ denote the chain above, which we want to show is small. Now $(*)$ can be non-zero only if B is relevant for $\overline{g}$ but there exists some face $\overline{g}_i = (g_0, \dots, \hat{g}_i, \dots, g_n)$ for which B is not relevant (otherwise just plug in the definition and check). This means that

$$diam_B(\pi_B(g_0) \cup \cdots \cup \pi_B(g_n)) \geq 2D; \text{ while}$$

$$diam_B(\pi_B(g_0) \cup \cdots \widehat{\pi_B(g_i)} \cdots \cup \pi_B(g_n)) < 2D.$$

Then when computing (∗) we obtain, for all such i, a sum of simplices in $\pi_B(g_0) \times \cdots \times \pi_B(\hat{g}_i) \times \cdots \times \pi_B(g_n)$, and by the above every such simplex is small.   □

9.5 Implications

At this point it is just a matter of a few computations to turn the intuitive idea of Subsection 9.2 into a real proof. The precise, quantitative statement is as follows. Recall the definition of the constant $K(\varphi)$ from Section 9.2: the supremum of φ over small simplices supported on H.

Theorem 9.18 [Frigerio et al., 2015, theorem 4.1] *For every $n \geq 1$ there exists a map*

$$\Theta^n \colon C_{alt}^n(H)^H \longrightarrow C_{alt}^n(G)^G$$

such that for all $\varphi \in C_{alt}^n(H)^H$ the following properties are satisfied:

1. $\sup\{\|\Theta^n(\varphi)(\overline{h}) - \varphi(\overline{h})\|_\infty \colon \overline{h} \in H^{n+1}\} \leq K(\varphi)$;
2. *If $n \geq 2$, then $\|\Theta^n(\varphi)\|_\infty \leq n(n+1)\|\varphi\|_\infty$;*
3. $\|\delta^n(\Theta^n(\varphi)) - \Theta^{n+1}(\delta^n(\varphi))\|_\infty \leq 2n(n+1)K(\varphi)$.

The first item says that $\Theta^n(\varphi)$ is (almost) an extension of φ; the second says that if φ is bounded, then so is $\Theta^n(\varphi)$; and the third says that Θ (almost) commutes with the coboundary operator: the last two items easily imply that Θ sends invariant quasicocycles to invariant quasicocycles.

As an immediate corollary, for all $n \geq 2$, the dimension of $EH_b^n(G)$ is bounded below by the dimension of $EH_b^n(H)$. Now if G has a non-degenerate hyperbolically embedded subgroup, then by Theorem 9.9 we may choose H to be of the form $F \times K$ for F a non-abelian free group and K a finite group. We already know that $H_b^2(F) = EH_b^2(F)$ is huge. It turns out that $H_b^3(F)$ has been computed as well, and is also huge [Soma, 1997a]. So we conclude:

Corollary 9.19 [Hull and Osin, 2013, corollary 1.7; Frigerio et al., 2015, corollary 5.5] *Let G be a group with a non-degenerate hyperbolically embedded subgroup (equivalently, an acylindrically hyperbolic group). Then $EH_b^2(G)$ and $EH_b^3(G)$ are uncountably dimensional.*

Example 9.20 By the theorems in Section 9.3, this applies to

1. Non-elementary hyperbolic and relatively hyperbolic groups;
2. The mapping class group $\mathrm{MCG}(\Sigma)$ of a closed surface of genus g with p punctures, provided $3g + p \geq 4$;
3. The outer automorphism group $Out(F)$ of a non-abelian free group;
4. Groups acting properly on a proper $CAT(0)$ space with a rank one isometry.

Beyond degree 3, nothing is known about $\mathrm{H}_b^n(F)$. But the same argument implies that if we were to find a non-trivial element here, this would imply non-vanishing for this much larger class of groups. The same can be said about *reduced* exact bounded cohomology classes, that is, those with non-vanishing Gromov seminorm [Frigerio et al., 2015, proposition 5.3].

For the sake of clarity, we restricted here to a single hyperbolically embedded subgroup and to the trivial G-module $\mathbb{R}$. But this construction works much more generally. Indeed, there is a notion of a *hyperbolically embedded family of subgroups* $(H_\lambda)_{\lambda \in \Lambda}$, which is defined the same way except taking a generating set X relative to $\cup_{\lambda \in \Lambda} H_\lambda^*$ [Dahmani et al., 2017, definition 4.25]. Then we can define an extension map with the same properties, whose domain is the direct sum of the spaces $\mathrm{C}_{alt}^n(H_\lambda)^{H_\lambda}$. Even better, we may replace $\mathbb{R}$ with any G-module V, and H_λ-invariant subspaces U_λ, the final result being a map

$$\Theta^n : \bigoplus_{\lambda \in \Lambda} \mathrm{C}_{alt}^n(H_\lambda; U_\lambda)^{H_\lambda} \longrightarrow \mathrm{C}_{alt}^n(G; V)^G.$$

Then the same result holds, and moreover $\Theta^n((\varphi_\lambda)_{\lambda \in \Lambda})(H_\mu^{n+1}) \subset U_\mu$ for any $\mu \in \Lambda$.

Being allowed to replace V with an invariant subspace can be very useful. For instance, take $V = \ell^2 G$ and $U = \ell^2 H$; this shows that if G is acylindrically hyperbolic, then $\mathrm{H}_b^2(G; \ell^2 G) \neq 0$ (the same result was obtained via a different method in Hamenstädt [2008]). This condition was introduced by Monod and Shalom [2006] as a cohomological definition of negative curvature in groups, for which a rich rigidity theory is developed.

10

Lie Groups and Symmetric Spaces

Anton Hase[*]

A Lie group is a group that is also a smooth manifold, such that multiplication and inversion are smooth. Lie groups appear as groups of smooth symmetries on manifolds, for example as isometry groups of Riemannian manifolds. This makes them both abundant and important. If we want to study non-discrete groups at all, we should have a look at Lie groups.

A symmetric space is a Riemannian manifold with isometric point reflections about every point. This class of spaces contains many of (at least) my favorite examples of Riemannian manifolds. The richness of the isometry group of a symmetric space gives us an association between symmetric spaces and Lie groups. Perhaps surprisingly, one can even classify symmetric spaces completely exploiting this association. From the viewpoint of group cohomology, the other direction of this association is the interesting one: we gain knowledge about a Lie group via the associated symmetric space.

In the first part of this chapter, we will sketch the association of symmetric spaces and Lie groups. In the second part, we will get back to the topic of (bounded) group cohomology. We present the definition of continuous and continuous bounded cohomology of a locally compact group. Then we see how a symmetric space provides us with a resolution for the continuous cohomology of the associated Lie group. A famous conjecture asks if the notions of continuous and continuous bounded cohomology coincide for connected semisimple Lie groups without compact factors and with finite center. We will report some evidence for the conjecture.

10.1 Symmetric Spaces

There are a number of textbooks about symmetric spaces, Lie groups, or both (see, for example, Loos [1969a,b], Wolf [2011], Knapp [2002]). In this part,

* Partially supported by ISF grant number 2019/19.

I will rely on the classic reference [Helgason, 1978]: definitions for symmetric spaces are taken from Chapter 4, for orthogonal symmetric algebras from Chapter 5, and for Lie groups and Lie algebras from Chapter 2. Some prior knowledge on Lie groups and Lie algebras is required to fully understand this part.

Definition 10.1 Let $p \in M$ be a point of a connected Riemannian manifold M. A diffeomorphism s_p defined on a neighbourhood N_p of p is a *geodesic symmetry with respect to p* if $s_p(\gamma(t)) = \gamma(-t)$ for every geodesic $\gamma \subset N_p$ satisfying $\gamma(0) = p$.

Definition 10.2 A connected Riemannian manifold M is a *symmetric space* if for every $p \in M$ the geodesic symmetry with respect to p can be extended to an isometry defined on all of M.

Example 10.3 Euclidean space is a symmetric space. The geodesic symmetry with respect to $p \in \mathbb{R}^n$ is the isometry $s_p(x) = 2p - x$.

Example 10.4 Real hyperbolic space is a symmetric space. The Minkowski bilinear form B on $\mathbb{R}^{n+1}$ is given by

$$B(x, y) = \sum_{i=1}^{n} x_i y_i - x_{n+1} y_{n+1}.$$

The hyperboloid model of real hyperbolic space $\mathbb{H}^n$ is

$$\mathbb{H}^n = \{x \in \mathbb{R}^{n+1} \mid B(x, x) = -1, x_{n+1} > 0\}$$

with the Riemannian metric on $T_p \mathbb{H}^n \cong \{x \in \mathbb{R}^{n+1} \mid B(p, x) = 0\}$ given by the restriction of B. The geodesic symmetry with respect to $p \in \mathbb{H}^n$ is the isometry $s_p(x) = -2B(x, p)p - x$.

Other prominent examples of symmetric spaces are spheres, real, complex, or quaternionic projective spaces and complex, quaternionic hyperbolic spaces. A more general class of Riemannian manifolds are locally symmetric spaces.

Definition 10.5 A connected Riemannian manifold M is a *locally symmetric space* if for every $p \in M$ there exists a neighborhood of p on which the geodesic symmetry with respect to p is an isometry.

In particular, any manifold that locally looks like a symmetric space is a locally symmetric space. This class includes, for example, flat and hyperbolic manifolds. (It is no coincidence that we see spaces of constant curvature here: a locally symmetric space can equivalently be defined as a connected Riemannian manifold, whose sectional curvature is invariant under parallel transport.) Even if the spaces one is interested in are only locally symmetric, there is a good reason to study symmetric spaces.

Theorem 10.6 [Helgason, 1978, corollary 5.7 in chapter IV] *The universal cover of a complete locally symmetric space is a symmetric space.*

Convinced that symmetric spaces are beautiful and important, we find a tool to study them:

Theorem 10.7 [Helgason, 1978, theorem 3.3 in chapter IV] *A symmetric space X is diffeomorphic to a homogeneous space G/K for a connected Lie group G and a compact subgroup K, which lies between the closed group K_σ of all fixpoints of an involution $\sigma : G \to G$ and the identity component of K_σ.*

Proof Let $G = \mathrm{Isom}^0(X)$. Then G is a connected Lie group and the action of G on X is smooth. (This follows from the Myers–Steenrod theorem; for a direct proof in this case, see Helgason [1978, pp. 205–208].) The symmetric space X is complete: Any geodesic γ can be extended using the geodesic symmetry s_p with respect to a point $p \in \gamma$. By the Hopf—Rinow theorem it follows that any two points $x, y \in X$ are connected by a geodesic γ. The geodesic symmetry with respect to the midpoint of γ sends x to y, so $\mathrm{Isom}(X)$ acts transitively on X. Since the action of $\mathrm{Isom}(X)$ on X is transitive and smooth, the G-orbit of any point is open. Since X is connected, all G-orbits are equal or, in other words, G acts transitively on X. Let K be the stabilizer of a point $p \in M$. Then K is compact and G/K is diffeomorphic to X.

Define $\sigma(g) = s_p \circ g \circ s_p$. Then $\sigma : G \to G$ is an involution. For any $k \in K$, the isometries k and $\sigma(k)$ have the same value and derivative at p, and therefore $k = \sigma(k)$ (see Helgason [1978, lemma 11.2 in chapter I]). This shows $K \subset K_\sigma$. On the other hand, K_σ commutes with s_p (using again Lemma 11.2), so it preserves the fixpoints of s_p. Since p is an isolated fixpoint of s_p, we get $K_\sigma^0 \subset K$. $\qquad\square$

Let $\mathfrak{g}$ and $\mathfrak{k}$ be the Lie algebras of G and K, respectively. Then $(d\sigma)_{eK} : \mathfrak{g} \to \mathfrak{g}$ is a Lie algebra involution, whose $+1$-eigenspace is $\mathfrak{k}$, since $K_\sigma^0 \subset K \subset K_\sigma$. If we denote the -1-eigenspace by $\mathfrak{p}$, we get the decomposition $\mathfrak{g} = \mathfrak{k} \oplus \mathfrak{p}$. Using the derivative of the projection $G \to G/K$, we can identify $\mathfrak{p}$ with the tangent space $T_{eK} G/K$. The G-invariant Riemannian metric on G/K corresponds to an $\mathrm{ad}(\mathfrak{k})$-invariant positive definite symmetric bilinear form Q on $\mathfrak{p}$. The symmetric space X is completely determined by G, K, σ, and Q.

Example 10.8 Real hyperbolic space is $\mathbb{H}^n = \mathrm{SO}^+(n, 1)/\mathrm{SO}(n)$. Here $\mathrm{SO}^+(n, 1)$ is the orthogonal group of the Minkowski quadratic form. The involution σ is given by $\sigma(g) = g^{-T}$. The Riemannian metric is a positive multiple of the Killing form on $\mathfrak{so}^+(n, 1)$ restricted to $\mathfrak{p}$.

Definition 10.9 Let $\mathfrak{g}$ be a real Lie algebra and $\theta \colon \mathfrak{g} \to \mathfrak{g}$ a Lie algebra involution. Let $\mathfrak{g} = \mathfrak{k} \oplus \mathfrak{p}$ be the decomposition of $\mathfrak{g}$ in eigenspaces of θ. If the connected Lie subgroup corresponding to $\mathrm{ad}(\mathfrak{k}) \subset \mathrm{ad}(\mathfrak{g})$ is compact, we call $(\mathfrak{g}, \theta)$ an *orthogonal symmetric algebra*.

We saw above that every symmetric space G/K gives us an orthogonal symmetric algebra $\mathfrak{g} = \mathfrak{k} \oplus \mathfrak{p}$ and an $\mathrm{ad}(\mathfrak{k})$-invariant positive definite symmetric bilinear form Q on $\mathfrak{p}$. On the other hand, we get a unique simply connected symmetric space given such a $\mathfrak{g} = \mathfrak{k} \oplus \mathfrak{p}$ and Q: take the simply connected Lie group G corresponding to $\mathfrak{g}$ and its subgroup K corresponding to $\mathfrak{k}$. Since G is simply connected, θ integrates to an involution $\sigma \colon G \to G$. Then G, K, σ, and Q determine a symmetric space. Right now, this is not a one-to-one correspondence even for simply connected symmetric spaces (different orthogonal symmetric algebras can give us the same symmetric space). However, one can get a one-to-one correspondence under additional assumptions. Continuing in this direction, one can completely classify symmetric spaces. We will not do so, as the classification of symmetric spaces is by far out of scope of this chapter. Instead, let us consider particular Lie groups that play a special role in (bounded) continuous group cohomology.

Definition 10.10 A connected Lie group is *semisimple* if all connected normal solvable subgroups are trivial. A connected, semisimple Lie group is *without compact factors* if all its connected normal compact subgroups are trivial.

Let G be a connected semisimple Lie group without compact factors. We associate to G the symmetric space G/K, where K is the maximal compact subgroup of G, with the Riemannian metric coming from the Killing form on $\mathfrak{g}$. The symmetric space G/K is simply connected, non-compact, has nonpositive curvature, and no Euclidean factor (see Helgason [1978, chapter VI]).

Definition 10.11 Let $(\mathfrak{g}, \theta)$ be an orthogonal symmetric algebra. Then the *dual* orthogonal symmetric algebra $(\mathfrak{g}^*, \theta^*)$ is given by $\mathfrak{g}^* = \mathfrak{k} \oplus i\mathfrak{p} \subset \mathfrak{g} \otimes_{\mathbb{C}} \mathbb{C}$ and $\theta^*(X + iY) = X - iY$ for $X \in \mathfrak{k}, Y \in \mathfrak{p}$.

Let G be a connected semisimple Lie group G without compact factors, and let G/K be its associated symmetric space. Let $(\mathfrak{g}, \theta)$ be the orthogonal symmetric algebra corresponding to G/K. Then its dual orthogonal symmetric algebra $(\mathfrak{g}^*, \theta^*)$ with the Killing form on $\mathfrak{g}^*$ provides us again with a simply connected symmetric space, which is called the *compact dual* of G/K. The compact dual symmetric space is compact, has non-negative curvature, and has no Euclidean factor. Using the one-to-one correspondence hinted at above, one obtains a duality between simply connected symmetric spaces with non-positive curvature and no Euclidean factor on one side, and simply connected compact symmetric spaces with non-negative curvature and no Euclidean factor on the other.

Example 10.12 The compact dual of real hyperbolic space $\mathbb{H}^n$ is the sphere $S^n = \mathrm{SO}(n+1)/\mathrm{SO}(n)$. Note that the sphere S^n and real projective space $\mathbb{R}P^n$ both correspond to the same orthogonal symmetric algebra. In this sense, they are both "dual" to hyperbolic space, but only the sphere is simply connected.

10.2 Continuous (Bounded) Cohomology of Lie Groups

In the introduction and the preceding chapters, the bounded cohomology of *discrete* groups has been defined and explored. Here we turn to the theory for *topological* groups. We recall the definition and the properties of continuous bounded cohomology. For more details, we refer the reader to the work of Burger and Monod [Monod, 2001; Burger and Monod, 2002].

Let G be a locally compact group. We can consider the space of *continuous functions* on $G^{\bullet+1}$, namely

$$\mathrm{C}_c^\bullet(G; \mathbb{R}) := \{f \colon G^{\bullet+1} \longrightarrow \mathbb{R} \mid f \text{ continuous}\},$$

with the natural G-action given by

$$(gf)(g_0, \ldots, g_\bullet) := f(g^{-1}g_0, \ldots, g^{-1}g_\bullet),$$

for every $f \in \mathrm{C}_c^\bullet(G; \mathbb{R})$ and $g, g_0, \cdots, g_\bullet \in G$.

There exists a natural norm of the space $\mathrm{C}_c^\bullet(G; \mathbb{R})$ given by

$$\|f\|_\infty := \sup\{|f(g_0, \ldots, g_\bullet)| \mid g_0, \ldots, g_\bullet \in G\},$$

and we say that f is bounded if its norm is finite. Denote by $\mathrm{C}_{cb}^\bullet(G; \mathbb{R})$ the subspace of *continuous bounded functions*.

The space of *continuous (bounded) G-invariant functions* is given by

$$\mathrm{C}_{c(b)}^\bullet(G; \mathbb{R})^G := \{f \in \mathrm{C}_{c(b)}^\bullet(G; \mathbb{R}) \mid gf = f \text{ for all } g \in G\}.$$

If we consider the *standard homogeneous coboundary operator*

$$\delta^\bullet : C_c^\bullet(G; \mathbb{R}) \longrightarrow C_c^{\bullet+1}(G; \mathbb{R}),$$

$$(\delta^\bullet f)(g_0, \ldots, g_{\bullet+1}) := \sum_{i=0}^{\bullet+1} (-1)^i f(g_0, \ldots, g_{i-1}, g_{i+1}, \ldots, g_{\bullet+1}),$$

we can notice that $\delta^\bullet$ preserves both the G-invariance and the boundedness.

Definition 10.13 [Monod, 2001, chapter III; Guichardet, 1980]   The *continuous bounded cohomology* with real coefficients $H_{cb}^\bullet(G; \mathbb{R})$ is the homology of the complex $(C_{cb}^\bullet(G; \mathbb{R})^G, \delta^\bullet)$. Similarly, the *continuous cohomology* $H_c^\bullet(G; \mathbb{R})$ is the homology of the complex $(C_c^\bullet(G; \mathbb{R})^G, \delta^\bullet)$.

Remark 10.14 1. For a discrete group G, the definitions above boil down to the usual definitions of (bounded) group cohomology: up to the additional condition of continuity of the functions, they are exactly the same as the ones for discrete groups presented in the introduction (see Definition 12 and before).

2. Continuous (bounded) cohomology can be defined with more general coefficients and computed by more than this resolution. For example, as for discrete groups, there is an inhomogeneous resolution (or *bar resolution*) that is particularly useful in low degrees (compare with Definition 14).

Notice that $H_{cb}^\bullet(G; \mathbb{R})$ admits a canonical seminormed structure obtained by considering the quotient seminorm

$$\|\alpha\| := \inf\{\|f\|_\infty \mid [f] = \alpha, \ f \in C_{cb}^\bullet(G; \mathbb{R})\},$$

with $\alpha \in H_{cb}^\bullet(G; \mathbb{R})$. Beyond the canonical seminorm, a way to measure the gap between the continuous cohomology and its bounded version relies on the *comparison map*

$$c_G^\bullet : H_{cb}^\bullet(G; \mathbb{R}) \longrightarrow H_c^\bullet(G; \mathbb{R}), \ [f]_b \longmapsto [f],$$

which is the map induced in cohomology by the canonical inclusion

$$\iota^\bullet : C_{cb}^\bullet(G; \mathbb{R}) \longrightarrow C_c^\bullet(G; \mathbb{R}).$$

The comparison map is in general neither injective nor surjective, since this is the case for discrete groups. But what can we say about the comparison map if G is a Lie group?

Conjecture 10.15 [Dupont, 1979; Monod, 2006a] If G is a connected semisimple Lie group without compact factors and with finite center, then the comparison map is an isomorphism.

Why do we concentrate on connected semisimple Lie groups without compact factors and with finite center here?

Proposition 10.16 [Monod, 2001, corollary 7.5.10] *Let G be a locally compact second countable topological group. Suppose $R \triangleleft G$ is a normal closed amenable subgroup. Then there is an isometric isomorphism $H^\bullet_{cb}(G) \cong H^\bullet_{cb}(G/R)$.*

Applying this proposition repeatedly, we get the following corollary:

Corollary 10.17 *Let G be a connected Lie group. Then there is a connected semisimple center-free Lie group without compact factors G' such that $H^\bullet_{cb}(G) \cong H^\bullet_{cb}(G')$.*

Proof Let $\tilde{G}$ be the universal covering group of G. Then the kernel of the projection $\tilde{G} \to G$ is in the center of $\tilde{G}$, so $H^\bullet_{cb}(G) \cong H^\bullet_{cb}(\tilde{G})$. Let R be the solvable radical of $\tilde{G}$, that is, the maximal connected normal solvable subgroup of $\tilde{G}$. If S is a maximal connected semisimple subgroup of $\tilde{G}$, then $\tilde{G} = R \rtimes S$ (which is called the *Levi–Malcev decomposition*). We get $H^\bullet_{cb}(G) \cong H^\bullet_{cb}(\tilde{G}/R) \cong H^\bullet_{cb}(S)$. The simply connected semisimple Lie group S is a direct product of simple Lie groups. If we factor out the compact factors of S, we get a simply connected semisimple Lie group S' without compact factors such that $H^\bullet_{cb}(G) \cong H^\bullet_{cb}(S')$. Factoring out the center of S', we arrive at G'.   □

This makes intuitive sense: abelian or compact factors give us non-negatively curved symmetric spaces, and bounded cohomology "detects" negative curvature. Continuous cohomology, on the other hand, is not invariant under amenable extensions, but only under finite extensions. Therefore the conjecture is as ambitious as possible. $\widetilde{SL}_2(\mathbb{R})$ is an example of a connected simple non-compact Lie group with infinite center for which the comparison map is not injective in degree two and not surjective in degree three.

How can we approach Conjecture 10.15? Finding resolutions to compute the continuous (bounded) cohomology of our groups is a good start.

Theorem 10.18 (See De la Cruz Mengual [2019, appendix A].) *Let G be a connected semisimple Lie group without compact factors and with finite center. Let X be the associated symmetric space and let X_c be its compact dual. Then*

$$H_c^\bullet(G) = \Omega^\bullet(X)^G = H^\bullet(X_c),$$

where $\Omega^k(X)^G$ denotes the G-invariant differential k-forms on X.

The cohomology of (irreducible) compact symmetric spaces is well studied (see Mimura and Toda [1991]), so this resolution is very useful. For continuous bounded cohomology, there is also a prominent resolution, which is explained in the next chapter (see Section 11.2). Unfortunately, the continuous bounded cohomology remains very hard to compute in higher degree. This is one of the reasons why Conjecture 10.15 is still wide open.

To finish this chapter, let us look at some of the evidence we have for Conjecture 10.15. In degrees 0 and 1, the conjecture is known and not difficult (see, for example, De la Cruz Mengual [2019, p. 28]). In degree 2 it was proven by Burger–Monod [Burger and Monod, 1999]. In degree 3 we still know the conjecture for a few groups: The conjecture holds for $SL_n(\mathbb{R})$ [Burger and Monod, 2002; Monod, 2004], $SL_n(\mathbb{C})$ [Goncharov, 1993; Bloch, 2000; Monod, 2004], and $SO^+(n, 1)$ [Pieters, 2018a]. It was also proven in degree 3 for $Sp_{2r}(\mathbb{C})$ for $r \geq 1$ [De la Cruz Mengual, 2019]. In degree 4 the conjecture is already only known for a single group, namely $SL_2(\mathbb{R})$ [Hartnick and Ott, 2015]. In general, for degree $n > 2$, the surjectivity of the comparison map is often known, while the injectivity remains mysterious. For more references concerning Conjecture 10.15, see De la Cruz Mengual [2019, Section 1.5] or the introduction of Ott [2019].

11

Continuous Bounded Cohomology, Representations, and Multiplicative Constants

Alessio Savini[*]

What is a multiplicative constant? Given a torsion-free lattice $\Gamma \leq G$ in a simple Lie group and a representation $\rho \colon \Gamma \to H$ into a locally compact group, a *multiplicative constant* is a numerical invariant that encodes important information about the conjugacy class of ρ.

A multiplicative constant is usually defined in terms of continuous bounded cohomology; its absolute value is bounded from above and its maximality implies the extendability of the representation to the ambient group.

We are going to give a general framework in which one can define the notion of a multiplicative constant. Then we are going to specialize it to several examples.

11.1 Continuous Bounded Cohomology

Let G be a locally compact group. We use the definitions of continuous (bounded) cohomology introduced in Section 10.2. Building on them, we briefly recall here some more properties of continuous bounded cohomology that we will exploit later. We again refer the reader to the work of Burger and Monod [Monod, 2001; Burger and Monod, 2002] for more details.

It should be clear to the reader who is confident in homological algebra that both continuous cohomology and continuous bounded cohomology are functorial. More precisely, given a continuous representation $\rho \colon G \to H$ into a locally compact group H, this induces a canonical map at the level of cochains

$$\mathrm{C}^{\bullet}_{c(b)}(\rho) \colon \mathrm{C}^{\bullet}_{c(b)}(H; \mathbb{R}) \longrightarrow \mathrm{C}^{\bullet}_{c(b)}(G; \mathbb{R}),$$

$$\mathrm{C}^{\bullet}_{c(b)}(\rho)(\psi)(g_0, \ldots, g_{\bullet}) := \psi(\rho(g_0), \ldots, \rho(g_{\bullet})).$$

[*] Supported by the Swiss National Science Foundation, grant no. 200020-192216.

108

It is easy to verify that the above is a well-defined map preserving boundedness. Additionally, it restricts to the subcomplexes of invariant vectors; hence it induces a map at the level of cohomology groups

$$\mathrm{H}^{\bullet}_{c(b)}(\rho)\colon \mathrm{H}^{\bullet}_{c(b)}(H;\mathbb{R}) \longrightarrow \mathrm{H}^{\bullet}_{c(b)}(G;\mathbb{R}).$$

Given a closed subgroup $L \leq G$, the map induced by the inclusion $i\colon L \hookrightarrow G$ is realized by the restriction map. Such a map admits a left inverse in the particular case when the quotient $L\backslash G$ admits a finite G-invariant measure (for instance, L is a lattice in G). In this case, one can define the map

$$\widehat{\mathrm{trans}}^{\bullet}_{L}\colon \mathrm{C}^{\bullet}_{cb}(L;\mathbb{R})^{L} \longrightarrow \mathrm{C}^{\bullet}_{cb}(G;\mathbb{R})^{G},$$

$$\widehat{\mathrm{trans}}^{\bullet}_{L}(\psi)(g_0,\cdots,g_{\bullet}) := \int_{L\backslash G} \psi(\overline{g}g_0,\cdots\overline{g}g_{\bullet})d\mu(\overline{g}),$$

where $\overline{g} \in L\backslash G$ is the equivalence class of g in the quotient and μ is the normalized G-invariant measure on $L\backslash G$.

Definition 11.1 The *cohomological transfer map* is the map induced in cohomology by $\widehat{\mathrm{trans}}^{\bullet}_{L}$, namely

$$\mathrm{trans}^{\bullet}_{L}\colon \mathrm{H}^{\bullet}_{cb}(L;\mathbb{R}) \longrightarrow \mathrm{H}^{\bullet}_{cb}(G;\mathbb{R}).$$

It is worth noticing that the existence of the transfer map relies crucially on the boundedness of cochains. To define a similar map in the context of continuous cohomology, one needs to require additional assumptions, such as the uniformity of the lattice [Monod, 2001, section 8.6].

11.2 Measurable and Essentially Bounded Functions

It may prove quite hard to compute the continuous bounded cohomology of a locally compact group G with only the help of Definition 10.13. To circumvent this problem, Burger and Monod in their works [Monod, 2001; Burger and Monod, 2002] exploited strong resolutions by relatively injective modules. This machinery would be too technical to be presented in this expository report, so we prefer to omit it. We instead focus our attention only on a specific resolution.

Consider a simple Lie group of non-compact type G (or a lattice inside such a Lie group) and consider a measure space (X, ν) on which G acts measurably, preserving the measure class of ν. We denote by

$$\mathcal{B}^{\infty}(X^{\bullet+1};\mathbb{R}) := \{f\colon X^{\bullet+1} \longrightarrow \mathbb{R} \mid f \text{ measurable and bounded}\}$$

the space of *bounded measurable functions* on $X^{\bullet+1}$. Similary, the space of *essentially bounded functions* on $X^{\bullet+1}$ is

$$L^{\infty}(X^{\bullet+1}, \mathbb{R}) := \mathcal{B}^{\infty}(X^{\bullet+1}; \mathbb{R})/\sim,$$

where $f \sim f'$ if they coincide on a set of full measure. With an abuse of nota-tion, we are going to refer to elements of L^{∞} by dropping the parenthesis and considering only a representative of the class.

We endow $\mathcal{B}^{\infty}(X^{\bullet+1}; \mathbb{R})$ (respectively $L^{\infty}(X^{\bullet+1}, \mathbb{R})$) with the usual G-action,

$$(gf)(\xi_0, \ldots, \xi_{\bullet}) := f(g^{-1}\xi_0, \ldots, g^{-1}\xi_{\bullet}),$$

for every $g \in G$ and almost every $\xi_0, \cdots, \xi_{\bullet} \in X$. We say that a function $f: X^{\bullet+1} \to \mathbb{R}$ is *alternating* if it holds

$$\mathrm{sign}(\sigma) f(\xi_0, \ldots, \xi_{\bullet}) = f(\xi_{\sigma(0)}, \ldots, f_{\sigma(\bullet)}),$$

where $\sigma \in S_{\bullet+1}$ is a permutation and $\mathrm{sign}(\sigma)$ denotes its signature (see Remark 13). Together with the standard homogeneous coboundary operator, we get two complexes: $(\mathcal{B}^{\infty}(X^{\bullet+1}; \mathbb{R}), \delta^{\bullet})$ and $(L^{\infty}(X^{\bullet+1}, \mathbb{R}), \delta^{\bullet})$.

Theorem 11.2 [Monod, 2001, theorem 7.5.3] *Let G be a simple Lie group of non-compact type. Consider $X = G/P$, where P is a minimal parabolic subgroup of G. Then it holds*

$$\mathrm{H}^{\bullet}(L^{\infty}(X^{\bullet+1}, \mathbb{R})^G) \cong \mathrm{H}^{\bullet}_{cb}(G; \mathbb{R}),$$

and the isomorphism preserves the standard seminormed structures. The same result holds if we substitute G with one of its lattices or we consider the subcomplex of alternating functions.

Example 11.3 Set $G = \mathrm{PSL}_2(\mathbb{R})$. Let $X = S^1$ be the circle and consider the stabilizer $P = \mathrm{Stab}_{\mathrm{PSL}_2(\mathbb{R})}(\xi_0)$ of a fixed point $\xi_0 \in S^1$. Since it holds that $S^1 \cong \mathrm{PSL}_2(\mathbb{R})/P$, then by Theorem 11.2 we get

$$\mathrm{H}^{\bullet}_{cb}(\mathrm{PSL}_2(\mathbb{R}); \mathbb{R}) \cong \mathrm{H}^{\bullet}(L^{\infty}((S^1)^{\bullet+1}, \mathbb{R})^{\mathrm{PSL}_2(\mathbb{R})}).$$

Example 11.4 Consider $G = \mathrm{Isom}^+(\mathbb{H}^3)$, for $n \geq 3$. Let $X = S^2$ be the sphere at infinity and consider $P = \mathrm{Stab}_{\mathrm{Isom}^+(\mathbb{H}^3)}(\xi_0)$ for a fixed $\xi_0 \in S^2$. Since we have $S^2 \cong \mathrm{Isom}^+(\mathbb{H}^3)/P$, then by Theorem 11.2 we get

$$\mathrm{H}^{\bullet}_{cb}(\mathrm{Isom}^+(\mathbb{H}^3); \mathbb{R}) \cong \mathrm{H}^{\bullet}(L^{\infty}((S^2)^{\bullet+1}, \mathbb{R})^{\mathrm{Isom}^+(\mathbb{H}^3)}).$$

To get the same isomorphism in higher dimension, we need to introduce a *twisted* action on the real coefficients of the cohomology groups. For more details, we refer the reader to Bucher et al. [2013, section 2].

Also the resolution of measurable bounded functions $(\mathcal{B}^\infty(X^{\bullet+1}; \mathbb{R}), \delta^\bullet)$ can be used to gain precious information about the bounded cohomology of G.

Theorem 11.5 [Burger and Iozzi, 2002, corollary 2.2] *Let (X, ν) be a meas-ure G-space where the measure class of ν is preserved by the G-action. Then there exists a canonical map*

$$\mathfrak{c}^\bullet \colon \mathrm{H}^\bullet(\mathcal{B}^\infty(X^{\bullet+1}; \mathbb{R})^G) \longrightarrow \mathrm{H}^\bullet_{cb}(G; \mathbb{R}).$$

Example 11.6 Let $G = \mathrm{Homeo}^+(S^1)$, the group of orientation-preserving homeomorphisms of the circle. For a fixed orientation, we can define a measurable function as follows:

$$\mathfrak{o} \colon (S^1)^3 \longrightarrow \mathbb{R},$$

$$\mathfrak{o}(\xi_0, \xi_1, \xi_2) := \begin{cases} 1/2 & \text{if } \xi_0, \xi_1, \xi_2 \text{ are positively oriented} \\ -1/2 & \text{if } \xi_0, \xi_1, \xi_2 \text{ are negatively oriented} \\ 0 & \text{otherwise.} \end{cases}$$

One can check that $\mathfrak{o}$ is an alternating, $\mathrm{Homeo}^+(S^1)$-invariant bounded measurable cocycle. It is called the *orientation cocycle*.

Using Theorem 11.5, we have a map

$$\mathfrak{c}^2 \colon \mathrm{H}^2(\mathcal{B}^\infty((S^1)^{\bullet+1}; \mathbb{R})^{\mathrm{Homeo}^+(S^1)}) \longrightarrow \mathrm{H}^2_b(\mathrm{Homeo}^+(S^1); \mathbb{R}),$$

where we considered $\mathrm{Homeo}^+(S^1)$ with the discrete topology. The class $\mathfrak{c}^2[\mathfrak{o}] \in \mathrm{H}^2_b(\mathrm{Homeo}^+(S^1); \mathbb{R})$ is non-trivial, and it is called the *bounded Euler class $e^b_\mathbb{R}$*.

Example 11.7 Set $G = \mathrm{Isom}^+(\mathbb{H}^3)$ and consider $X = S^2$. We define the *volume function* as

$$\mathrm{Vol}_3 \colon (S^2)^4 \longrightarrow \mathbb{R}, \quad \mathrm{Vol}_3(\xi_0, \dots, \xi_3) = \int_{\Delta(\xi_0, \dots, \xi_3)} \omega,$$

where $\omega \in \Omega^3(\mathbb{H}^3)^{\mathrm{Isom}^+(\mathbb{H}^3)}$ is the volume form and $\Delta(\xi_0, \dots, \xi_3)$ is the hyperbolic convex hull of the 4-tuple. The function Vol_3 is alternating, $\mathrm{Isom}^+(\mathbb{H}^3)$-invariant, and measurable. Moreover, its absolute value is bounded from above by the volume v_3 of a regular ideal tetrahedron (see Section 2.1).

Again by Theorem 11.5 we have a map

$$\mathfrak{c}^3 \colon \mathrm{H}^3(\mathcal{B}^\infty((S^2)^{\bullet+1}; \mathbb{R})^{\mathrm{Isom}^+(\mathbb{H}^3)}) \longrightarrow \mathrm{H}^3_{cb}(\mathrm{Isom}^+(\mathbb{H}^3); \mathbb{R}),$$

which, in this case, is induced by the projection $\mathcal{B}^\infty((S^2)^{\bullet+1}; \mathbb{R}) \rightarrow L^\infty((S^2)^{\bullet+1}, \mathbb{R})$. The class $\Theta^b_3 := \mathfrak{c}^3[\mathrm{Vol}_3]$ is non-trivial and it is called the *Volume class*. The volume class Θ_3^b is a generator for the group $\mathrm{H}_{cb}^3(\mathrm{Isom}^+(\mathbb{H}^3); \mathbb{R})$.

Example 11.8 Let G be a simple *Hermitian* Lie group, that is, whose associated symmetric space $\mathcal{X}$ admits a G-invariant complex structure $\mathcal{J}$ which is compatible with the Riemannian metric $\langle \cdot, \cdot \rangle$. We can define the *Kähler form* $\omega \in \Omega^2(\mathcal{X})^G$ as

$$\omega_a(X, Y) := \langle \mathcal{J}_a X, Y \rangle_a,$$

for every $a \in \mathcal{X}$ and $X, Y \in T_a \mathcal{X}$. For any triple of *distinct* points $a, b, c \in \mathcal{X}$, we can define

$$c_\omega(a, b, c) := \frac{1}{2\pi} \int_{\Delta(a,b,c)} \omega,$$

where $\Delta(a, b, c)$ is a triangle with geodesic sides. The function c_ω is bounded, alternating, and satisfies the cocycle condition (since ω is closed). Additionally, it can be extended measurably to the Shilov boundary $\check{\mathcal{S}}_\mathcal{X}$ of the symmetric space $\mathcal{X}$ [Clerc and Ørsted, 2003], producing a bounded measurable cocycle

$$\beta_\mathcal{X} : (\check{\mathcal{S}}_\mathcal{X})^3 \longrightarrow \mathbb{R},$$

called the *Bergmann cocycle*. Recalling that the Shilov boundary is a homogeneous G-space identified with the quotient of G by a suitable *maximal parabolic* subgroup, thanks to Theorem 11.5 we get a map

$$\mathfrak{c}^2 : \mathrm{H}^2(\mathcal{B}^\infty((\check{\mathcal{S}}_\mathcal{X})^{\bullet+1}; \mathbb{R})^G) \longrightarrow \mathrm{H}_{cb}^2(G; \mathbb{R}).$$

The class defined by $\kappa_G^b := \mathfrak{c}^2[\beta_\mathcal{X}]$ is non-trivial and it is called the *bounded Kähler class*. It is a generator for the group $\mathrm{H}_{cb}^2(G; \mathbb{R})$, which is one-dimensional.

Notice that in the particular case of $G = \mathrm{PU}(1, 1) \cong \mathrm{PSL}_2(\mathbb{R})$, the Shilov boundary is the circle S^1 and the Bergmann cocycle boils down to the *orientation cocycle* of Example 11.6.

We conclude by showing how one can efficiently implement the pullback map in bounded cohomology. Let $\rho : G \to H$ be a continuous representation into a locally compact group H and let (Y, θ) be a measure space on which H acts by preserving the measure class of θ.

Definition 11.9 Let P be a *minimal parabolic subgroup* of G. A *boundary map* is a measurable map $\varphi : G/P \to Y$ which is ρ-*equivariant*, that is,

$$\varphi(g\xi) = \rho(g)\varphi(\xi),$$

for almost every $g \in G$ and almost every $\xi \in G/P$. On G we are considering the Haar measurable structure.

A boundary map is a powerful tool to compute an explicit representative of a pullback map. Indeed, one has the following:

Theorem 11.10 [Burger and Iozzi, 2002, corollary 2.7] *Let $\rho \colon G \to H$ be a continuous representation. Suppose that ρ admits a boundary map $\varphi \colon G/P \to Y$. Let $\alpha \in \mathrm{H}_{cb}^{\bullet}(H; \mathbb{R})$ be a bounded class which is represented by a measurable function $\psi \in \mathcal{B}^{\infty}(Y^{\bullet+1}; \mathbb{R})^{H}$, that is, $\mathrm{c}^{\bullet}([\psi]) = \alpha$. Then the pullback class $\mathrm{H}_{cb}^{\bullet}(\rho)(\alpha)$ admits as a canonical representative*

$$\mathrm{C}^{\bullet}(\varphi)(\psi)(\xi_0, \ldots, \xi_{\bullet}) := \psi(\varphi(\xi_0), \ldots, \varphi(\xi_{\bullet})) \in L^{\infty}((G/P)^{\bullet+1}, \mathbb{R})^{G}.$$

11.3 Multiplicative Constants

Having introduced all the tools that we needed, we are finally ready to state the main result about multiplicative constants. The result is quite technical, so we are going to give some examples later for the sake of clearness.

Theorem 11.11 [Burger and Iozzi, 2009] *Let $\Gamma \leq G$ be a lattice in a simple Lie group of non-compact type and let $\rho \colon G \to H$ be a continuous representation into a locally compact group. Let $P \leq G$ be a minimal parabolic subgroup and let (Y, θ) be a measurable H-space where the measure class of θ is preserved by H. Suppose that there exists a boundary map $\varphi \colon G/P \to Y$. Consider $\psi' \in \mathcal{B}^{\infty}(Y^{\bullet+1}; \mathbb{R})^{H}$ an alternating H-invariant bounded measurable cocycle. Fix an alternating G-invariant essentially bounded cocycle $\psi \in L^{\infty}((G/P)^{\bullet+1}, \mathbb{R})^{G}$ and denote by $\Psi \in \mathrm{H}_{cb}^{\bullet}(G; \mathbb{R})$ its associated class. If $\mathrm{H}_{cb}^{\bullet}(G; \mathbb{R}) \cong \mathbb{R} \cdot \Psi$, then there exists a real number $\lambda_{\psi', \psi}(\rho) \in \mathbb{R}$ such that*

$$\int_{\Gamma \backslash G} \psi'(\varphi(\overline{g}\xi_0), \ldots, \varphi(\overline{g}\xi_{\bullet})) d\mu(\overline{g}) = \lambda_{\psi', \psi}(\rho) \psi(\xi_0, \ldots, \xi_{\bullet}) + \text{coboundary}$$

$$(11.1)$$

for almost every $\xi_0, \ldots, \xi_{\bullet} \in G/P$.

Definition 11.12 In the situation of Theorem 11.11, we call the number $\lambda_{\psi', \psi}(\rho)$ the *multiplicative constant* associated to the representation ρ and the cocycles ψ', ψ.

Suppose now that in Equation (11.1) no coboundary terms appear. Then it is easy to verify that

$$|\lambda_{\psi',\psi}(\rho)| \leq \frac{\|\psi'\|_\infty}{\|\psi\|_\infty}. \tag{11.2}$$

This leads naturally to the following:

Definition 11.13 In the situation of Theorem 11.11, suppose that no coboundary terms appear in Equation (11.1). We say that a representation is *maximal* if it satisfies

$$|\lambda_{\psi',\psi}(\rho)| = \frac{\|\psi'\|_\infty}{\|\psi\|_\infty}.$$

Remark 11.14 There are many situations in which no coboundary terms appear in Equation (11.1) (see Burger and Iozzi [2009, remark 3.2]). For instance, this happens when Γ admits an ergodic action on the product $(G/P)^\bullet$ [Burger and Mozes, 1996].

11.4 Examples and Applications

11.4.1 The Euler Invariant of a Representation

Let $G = \mathrm{PSL}_2(\mathbb{R})$ and consider a torsion-free uniform lattice $\Gamma \leq G$. Consider $H = \mathrm{Homeo}^+(S^1)$ and let $Y = S^1$ endowed with the usual Lebesgue measure. Denote by $\Sigma = \Gamma\backslash\mathbb{H}^2$ the closed hyperbolic surface associated with Γ. Recall that by Gromov's mapping theorem 1.2 [Gromov, 1982], we have an isomorphism $\mathrm{H}_b^2(\Gamma; \mathbb{R}) \cong \mathrm{H}_b^2(\Sigma; \mathbb{R})$.

Definition 11.15 The *Euler invariant* associated to the representation $\rho \colon \Gamma \to H$ is the number

$$\mathrm{eu}(\rho) := \langle c_\Sigma^2 \circ \mathrm{H}_b^2(\rho)(e_{\mathbb{R}}^b), [\Sigma]\rangle,$$

where c_Σ^2 is the comparison map relative to the surface Σ, $[\Sigma] \in \mathrm{H}_2(\Sigma; \mathbb{R})$ is a fixed fundamental class, and $\langle \cdot, \cdot \rangle$ denotes the Kronecker pairing (see Equation (E1) from the Introduction).

Since $\mathrm{H}_b^2(\rho)$ depends only on the conjugacy class of ρ, the same holds for $\mathrm{eu}(\rho)$. We are going to understand the relation between the Euler invariant and the notion of multiplicative constant. Suppose that there exists a *boundary map* $\varphi \colon S^1 \to S^1$ for ρ (for instance, ρ is *non-elementary*). In the notation of Theorem 11.11, we fix $\psi' = \psi = \mathrm{o}$, where o is the orientation cocycle of Example 11.6.

As mentioned at the end of Example 11.8, the Euler class $e_{\mathbb{R}}^b \in$ $H_{cb}^2(\mathrm{PSL}_2(\mathbb{R}); \mathbb{R})$ is a generator. Hence the hypotheses of Theorem 11.11 are satisfied and we get that there exists a real number $\lambda_{o,o}(\rho)$ that satisfies

$$\int_{\Gamma/\mathrm{PSL}_2(\mathbb{R})} o(\varphi(\overline{g}\xi_0), \varphi(\overline{g}\xi_1), \varphi(\overline{g}\xi_2)) d\mu(\overline{g}) = \lambda_{o,o}(\rho)o(\xi_0, \xi_1, \xi_2),$$

for almost every $\xi_0, \xi_1, \xi_2 \in S^1$. We did not write any coboundary term in the above formula since it is known that Γ acts doubly ergodically on S^1 (see Remark 11.14). As a consequence, if we specialize Equation (11.2) to this particular case, we obtain

$$|\lambda_{o,o}(\rho)| \leq 1. \tag{11.3}$$

As shown by Iozzi [Iozzi, 2002], it holds that

$$\mathrm{eu}(\rho) = \lambda_{o,o}(\rho) \cdot \chi(\Sigma),$$

and thanks to Equation (11.3), it follows immediately that

$$|\mathrm{eu}(\rho)| \leq |\chi(\Sigma)|.$$

Those representations which attain the maximum are called *maximal* (see Definition 11.13). One has the following:

Theorem 11.16 [Matsumoto, 1987; Iozzi, 2002] *Let* $\Gamma \leq \mathrm{PSL}_2(\mathbb{R})$ *be a torsion-free uniform lattice and let* $\rho\colon \Gamma \to \mathrm{Homeo}^+(S^1)$. *Then* ρ *is maximal if and only if it is semiconjugated to the standard lattice embedding* $\Gamma \to \mathrm{PSL}_2(\mathbb{R})$.

11.4.2 The Volume of a Representation and the Mostow Rigidity Theorem

Let $G = H = \mathrm{Isom}^+(\mathbb{H}^3)$ and consider a torsion-free uniform lattice $\Gamma \leq G$ (the theorem holds also for non-uniform ones, but the notation is easier in the compact case). Denote by $M = \Gamma\backslash\mathbb{H}^3$ the closed hyperbolic 3-manifold associated with Γ.

Definition 11.17 The *volume* of the representation $\rho\colon \Gamma \to H$ is the number

$$\mathrm{Vol}(\rho) := \langle c_M^3 \circ H_b^3(\rho)(\Theta_b^3), [M] \rangle,$$

where c_M^3 is the comparison map associated with M, $[M] \in H_3(M; \mathbb{R})$ is a fixed fundamental class, and $\langle \cdot, \cdot \rangle$ is the Kronecker pairing.

Notice that we exploited tacitly Gromov's mapping theorem 1.2 by identifying $H^3_b(\Gamma; \mathbb{R})$ with $H^3_b(M; \mathbb{R})$. Also in this case, the volume is constant along the conjugacy class of ρ and it is related to the notion of multiplicative constant. By fixing $Y = S^2$, we are going to suppose that ρ admits a *boundary map* $\varphi: S^2 \to S^2$ (for instance, ρ is *non-elementary*). Following the notation of Theorem 11.11, we set $\psi' = \psi = \mathrm{Vol}_3$, where Vol_3 is the volume function of Example 11.7.

Since the volume class $\Theta^b_3 \in H^3(\mathrm{Isom}^+(\mathbb{H}^3); \mathbb{R})$ is a generator, we can again apply Theorem 11.11. Hence there exists $\lambda_{\mathrm{Vol}_3,\mathrm{Vol}_3}(\rho)$ such that

$$\int_{\Gamma \backslash \mathrm{Isom}^+(\mathbb{H}^3)} \mathrm{Vol}_3(\varphi(\overline{g}\xi_0), \ldots, \varphi(\overline{g}\xi_3)) d\mu(\overline{g}) = \lambda_{\mathrm{Vol}_3,\mathrm{Vol}_3}(\rho) \mathrm{Vol}(\xi_0, \ldots, \xi_3),$$

for almost every $\xi_0, \cdots, \xi_3 \in S^2$. Since Γ acts doubly ergodically on S^2, there are no coboundary terms (Remark 11.14), and using Equation 11.2, we deduce

$$|\lambda_{\mathrm{Vol}_3,\mathrm{Vol}_3}(\rho)| \le 1. \tag{11.4}$$

Bucher, Burger, and Iozzi [Bucher et al., 2013] proved that

$$\mathrm{Vol}(\rho) = \lambda_{\mathrm{Vol}_3,\mathrm{Vol}_3}(\rho) \cdot \mathrm{Vol}(M),$$

which implies, together with Equation (11.4), the following estimate:

$$|\mathrm{Vol}(\rho)| \le \mathrm{Vol}(M).$$

A representation attaining the maximum is called *maximal* and for that we have a sort of Mostow rigidity theorem.

Theorem 11.18 [Bucher et al., 2013]   *Let* $\Gamma \le \mathrm{Isom}^+(\mathbb{H}^3)$ *a torsion-free uniform lattice and let* $\rho: \Gamma \to \mathrm{Isom}^+(\mathbb{H}^3)$ *be a representation. Then* $\mathrm{Vol}(\rho) = \mathrm{Vol}(M)$ *if and only if* ρ *is conjugated to the standard lattice embedding.*

11.4.3 Toledo Invariant and Higher Teichmüller Theory

Let $\Gamma \le \mathrm{PSL}_2(\mathbb{R})$ be a torsion-free uniform lattice and let H be a Hermitian Lie group. Consider the closed hyperbolic surface $\Sigma = \Gamma \backslash \mathbb{H}^2$ associated with Γ.

Definition 11.19 The *Toledo invariant* of a representation $\rho: \Gamma \to H$ is the real number

$$T_b(\rho) := \langle c^2_\Sigma \circ H^2_b(\rho)(\kappa^b_H), [\Sigma] \rangle,$$

where the notation we used is the same as in Definition 11.15.

Let $\mathcal{X}$ be the symmetric space associated with H and let $Y = \check{\mathcal{S}}_\mathcal{X}$ be the Shilov boundary. We are going to suppose that ρ admits a boundary map $\varphi \colon S^1 \to \check{\mathcal{S}}_\mathcal{X}$. Burger and Iozzi [2004] proved that a boundary map exists for *Zariski dense* representations. Following the notation of Theorem 11.11, we set $\psi' = \beta_\mathcal{X}$ and $\psi = \mathfrak{o}$, where $\beta_\mathcal{X}$ is the Bergmann cocycle of Example 11.8.

Since the bounded Kähler class $\kappa_H^b \in \mathrm{H}_{cb}^2(H; \mathbb{R})$ is a generator, Theorem 11.11 guarantees the existence of a number $\lambda_{\beta_\mathcal{X}, \mathfrak{o}}(\rho)$ such that

$$\int_{\Gamma \backslash \mathrm{PSL}_2(\mathbb{R})} \beta_\mathcal{X}(\varphi(\overline{g}\xi_0), \varphi(\overline{g}\xi_1), \varphi(\overline{g}\xi_2)) d\mu(\overline{g}) = \lambda_{\beta_\mathcal{X}, \mathfrak{o}}(\rho)\mathfrak{o}(\xi_0, \xi_1, \xi_2),$$

for almost every $\xi_0, \xi_1, \xi_2 \in S^1$. The ubiquitous double ergodic action of Γ on S^1 allows one to omit the coboundary term (Remark 11.14). Equation (11.2) implies that

$$|\lambda_{\beta_\mathcal{X}, \mathfrak{o}}| \leq \mathrm{rank}(\mathcal{X}), \tag{11.5}$$

where $\mathrm{rank}(\mathcal{X})$ denotes the real rank of $\mathcal{X}$ (i.e., the dimension of a maximal flat). Burger, Iozzi, and Wienhard [Burger et al., 2010] related the Toledo invariant to the multiplicative constant $\lambda_{\beta_\mathcal{X}, \mathfrak{o}}$ as follows:

$$\mathrm{T}_b(\rho) = \lambda_{\beta_\mathcal{X}, \mathfrak{o}} \cdot \chi(\Sigma).$$

Using the equation above jointly with Equation (11.5), we immediately argue that

$$|\mathrm{T}_b(\rho)| \leq \mathrm{rank}(\mathcal{X}) \cdot |\chi(\Sigma)|.$$

Also in this case, a representation attaining the maximum is called *maximal representation*. Maximal representations are crucial in the study of higher Teichmüller theory and they were completely characterized by Burger, Iozzi, and Wienhard. Before giving the statement, recall that a Hermitian symmetric space is of *tube type* if it can be written as $V + i\Omega$, where V is a real vector space and Ω is an open cone in V.

Theorem 11.20 [Burger et al., 2010] *Let $\Gamma \leq \mathrm{PSL}_2(\mathbb{R})$ be a torsion-free uniform lattice and let $\rho \colon \Gamma \to H$ be a representation into a Hermitian Lie group. Denote by $\mathbf{L} = \overline{\rho(\Gamma)}^Z$ the Zariski closure of the image and let $L = \mathbf{L}(\mathbb{R})^\circ$ be the connected component of the real points. If ρ is maximal, then $\mathbf{L}$ is reductive, the centralizer $Z_H(L)$ is compact, and the symmetric space associated with L is of tube type. Moreover, ρ is discrete and injective.*

12

The Proportionality Principle via Bounded Cohomology

Filippo Baroni

In Chapter 2, we saw a geometric proof of Gromov's proportionality principle (Theorem 5 in the introduction), a classical result that establishes a connection between the topology of a closed Riemannian manifold (namely, its simplicial volume) and its geometric structure. In this chapter, we will present a more "algebraic" proof, relying heavily on the isometric isomorphism between singular and continuous cohomology. Our strategy involves finding a specific continuous cohomology class (the *volume coclass*) whose ℓ^∞-seminorm will determine the proportionality constant between Riemannian and simplicial volume. Finally, as an application, we will compute this constant for hyperbolic manifolds, showing that their simplicial volume is always non-vanishing.

For simplicity, we will also assume that M is orientable; the non-orientable case can be easily dealt with by taking the orientable double covering of M and exploiting the multiplicativity of simplicial volume for finite coverings.

The aim of this chapter is to prove a generalization of the proportionality principle presented in Chapter 2 (Theorem 2.1).

Theorem 12.1 (Proportionality principle [Theorem 5]) *Let M be a closed Riemannian manifold. Then the ratio between its simplicial and Riemannian volume only depends on the isometry type of the universal Riemannian covering of M.*

We will present a proof of this fact by means of (continuous) bounded cohomology, while also finding a somewhat explicit expression for the proportionality constant. This approach was described in details by Frigerio first in Frigerio [2011] and then in Frigerio [2017, chapter 8]. We will follow here the latter reference. In particular, we will work under the hypothesis that M is non-positively-curved, which will greatly simplify our exposition by providing us with a very convenient straightening procedure. This restriction, however, is not necessary: for a fully general proof of the proportionality principle, using a similar strategy but more advanced techniques, we refer the reader to Frigerio [2011].

12.1 Straightening in Non-positive Curvature

The proof presented in Chapter 2 relies on a straightening operator that turns arbitrary singular simplices of M into smooth ones. In this section, we will discuss a similar procedure that generalizes well to non-hyperbolic manifolds.

12.1.1 Straight Simplices

Let $\widetilde{M}$ be a simply connected oriented complete Riemannian n-manifold with non-positive sectional curvature: the Cartan–Hadamard theorem implies that every pair of points in $\widetilde{M}$ is connected by a unique geodesic. We will now describe a straightening procedure for simplices in $\widetilde{M}$.

Definition 12.2 [Frigerio, 2017, section 8.4] Let $x_0, \ldots, x_k$ be points in $\widetilde{M}$. The *straight simplex* with vertices $x_0, \ldots, x_k$ is the $(k + 1)$-simplex $[x_0, \ldots, x_k]: \Delta^k \to \widetilde{M}$ defined inductively as follows:

- for $k = 0$, the straight simplex $[x_0]$ is simply the point x_0;
- for $k > 0$, the straight simplex $[x_0, \ldots, x_k]$ is the "geodesic cone" with vertex x_k and base $[x_0, \ldots, x_{k-1}]$; more precisely, for every $z \in \Delta^{k-1} \subseteq \Delta^k$, the restriction of $[x_0, \ldots, x_k]$ to the segment with endpoints[1] z and e_k is the constant-speed parametrization of the geodesic in $\widetilde{M}$ connecting $[x_0, \ldots, x_{k-1}](z)$ and x_k.

Since geodesics depend continuously (smoothly) on their endpoints, it is easy to see that $[x_0, \ldots, x_k]$ is indeed a continuous (smooth) map from Δ^k to $\widetilde{M}$.

12.1.2 The Straightening Chain Map

We can exploit the procedure we have described to define a *straightening chain map*. For every k-simplex s set

$$\widetilde{\mathrm{str}}_k(s) := [s(e_0), \ldots, s(e_k)];$$

if we extend by linearity, we get a map $\widetilde{\mathrm{str}}_k \colon C_k(\widetilde{M}) \to C_k(\widetilde{M})$ for every k.

---

[1] Here e_i denotes the ith vertex of the standard simplex $\Delta^k \subseteq \mathbb{R}^{k+1}$.

Let G be the group of orientation-preserving isometries of $\widetilde{M}$. Recall that the chain modules[2] $C_\bullet(\widetilde{M})$ are naturally endowed with a left G-action. We have the following:

Proposition 12.3 [Frigerio, 2017, proposition 8.11] *The maps*

$$\widetilde{\mathrm{str}}_k \colon C_k(\widetilde{M}) \longrightarrow C_k(\widetilde{M})$$

enjoy these properties:

1. $\widetilde{\mathrm{str}}_k(s)$ *is a smooth simplex for every k-simplex s;*
2. $\partial \circ \widetilde{\mathrm{str}}_k = \widetilde{\mathrm{str}}_{k-1} \circ \partial$ *(where ∂ is the boundary map); therefore,*

$$\widetilde{\mathrm{str}}_\bullet \colon C_\bullet(\widetilde{M}) \longrightarrow C_\bullet(\widetilde{M})$$

is a chain map;
3. *the chain map $\widetilde{\mathrm{str}}_\bullet$ is G-invariant;*
4. *the chain map $\widetilde{\mathrm{str}}_\bullet$ is homotopic to the identity through a G-invariant homotopy, sending smooth simplices to a linear combination of smooth simplices.*

This allows us to extend our straightening procedure to possibly non-simply-connected manifolds. Let M be an oriented closed connected Riemannian n-manifold with non-positive sectional curvature, and let $p\colon \widetilde{M} \to M$ be its universal covering. Denote by Γ the subgroup of G given by the deck transformations of p. We will maintain this notation until the end of this chapter. Proposition 12.3 immediately gives the following:

Corollary 12.4 [Frigerio, 2017, p. 113] *There exists a chain map* $\mathrm{str}_\bullet \colon C_\bullet(M) \to C_\bullet(M)$ *such that*

1. $\mathrm{str}_k(s)$ *is a smooth simplex for every k-simplex s;*
2. $\mathrm{str}_\bullet$ *is homotopic to the identity through a homotopy sending smooth simplices to linear combinations of smooth simplices.*

12.2 Duality and the Volume Coclass

In this section, we will explore the strong connection between simplicial volume and bounded cohomology, providing a proof of the duality principle (Proposition 10). We will then construct a cohomology class (the so-called *volume coclass*) whose ℓ^∞-seminorm is exactly equal to the ratio between the Riemannian and the simplicial volume of M.

---

[2] We adopt the usual convention of omitting the coefficient group when dealing with homology and cohomology with real coefficients.

12.2.1 Duality

We first introduce a classical duality result relating the ℓ^1 and ℓ^∞-seminorms [Frigerio, 2017, section 6].

Proposition 12.5 [Gromov, 1982] *Let X be a topological space. For every homology class $\alpha \in H_k(X)$, the following holds:*

$$\|\alpha\|_1 = \max \left\{ |\langle \beta, \alpha \rangle| : \beta \in H^k(X), \|\beta\|_\infty \leq 1 \right\},$$

where $\langle -, - \rangle$ denotes the Kronecker pairing (see Equation (E1) from the Introduction).

Proof The inequality ($\geq$) follows immediately from the fact that

$$|\langle \beta, \alpha \rangle| \leq \|\beta\|_\infty \cdot \|\alpha\|_1 \leq \|\alpha\|_1$$

for every β whose seminorm is not greater than 1. For the opposite inequality, let $a \in C_k(X)$ be a cycle representing α. Let $B \subseteq C_k(X)$ be the subspace of the boundaries. The Hahn–Banach theorem guarantees the existence of a functional $\varphi \colon C_k(X) \to \mathbb{R}$ of norm at most 1 that vanishes on B and satisfies

$$\varphi(a) = \inf \left\{ \|a - b\|_1 : b \in B \right\} = \|\alpha\|_1,$$

proving the inequality ($\leq$). $\qquad\square$

Let $[M] \in H_n(M)$ be the (real) fundamental class of M, that is, the image of the integral fundamental class $[M]_\mathbb{Z}$ under the change of coefficients map $H_n(M; \mathbb{Z}) \to H_n(M)$. By Poincaré duality, there exists a unique cohomology class $[M]^* \in H^n(M)$ such that $\langle [M]^*, [M] \rangle = 1$; we call it the *fundamental coclass* of M.

We denote by $\|M\|$ the simplicial volume of M, that is, the ℓ^1-seminorm of its fundamental class $[M]$.

Proposition 12.6 (Duality principle [Proposition 10]) *We have*

$$\|M\| = \left(\|[M]^*\|_\infty \right)^{-1}.$$

Here we agree that $\infty^{-1} = 0$.

Proof By Proposition 12.5, we have

$$\|M\| = \max \left\{ \langle \beta, [M] \rangle : \beta \in H^n(M), \|\beta\|_\infty \leq 1 \right\}.$$

If $\|[M]^*\|_\infty = \infty$, then every non-trivial cohomology class is unbounded; therefore the right-hand side vanishes. Otherwise $[M]^*/\|[M]^*\|_\infty$ is the only coclass of seminorm 1; therefore

$$\|M\| = \frac{1}{\|[M]^*\|_\infty} \cdot \langle [M]^*, [M] \rangle = \frac{1}{\|[M]^*\|_\infty}. \qquad \square$$

12.2.2 The Volume Cocycle

We will now define an explicit cocycle that will turn out to be very useful in our proof of the proportionality principle. For every n-simplex s of M, let

$$\mathrm{Vol}_M(s) := \int_{\mathrm{str}_n(s)} \omega,$$

where ω denotes the Riemannian volume form of M. Extending by linearity, we get a cochain $\mathrm{Vol}_M \in C^n(M)$, which we name the *volume cochain* of M. Actually, it is easy to see that Vol_M is a cocycle. In fact, for every $c \in C_{n+1}(M)$, we have

$$\int_{\mathrm{str}_n(dc)} \omega = \int_{d\,\mathrm{str}_{n+1}(c)} \omega = \int_{\mathrm{str}_n(c)} d\omega = 0,$$

where we used the fact that $\mathrm{str}_\bullet$ is a chain map and Stokes' theorem.

12.2.3 The Volume Coclass

The following proposition explains our interest in the volume cocycle.

Proposition 12.7 [Frigerio, 2017, lemma 8.12] *Let* $[\mathrm{Vol}_M] \in H^n(M)$ *be the cohomology class of the volume cocycle. Then the following holds:*

$$[\mathrm{Vol}_M] = \mathrm{vol}(M) \cdot [M]^*,$$

where $\mathrm{vol}(M)$ *denotes the Riemannian volume of* M.

Proof Let $c \in C_n(M)$ be the cycle associated with a smooth triangulation of M; it is well known that c is a representative of the fundamental class of M. Since $H^n(M)$ is one-dimensional, we have $[\mathrm{Vol}_M] = \mathrm{Vol}_M(c) \cdot [M]^*$ (recall that $\langle [M]^*, [c] \rangle = 1$). But

$$\mathrm{Vol}_M(c) = \int_{\mathrm{str}_n(c)} \omega = \int_c \omega + \int_{\mathrm{str}_n(c)-c} \omega = \int_c \omega = \mathrm{vol}(M),$$

where the integral over $\mathrm{str}_n(c) - c$ vanishes due to Stokes' theorem, since by Corollary 12.4 the chain $\mathrm{str}_n(c) - c$ is a linear combination of smooth boundaries. $\qquad \square$

Corollary 12.8 *We have*

$$\frac{\|M\|}{\mathrm{vol}(M)} = \frac{1}{\|[\mathrm{Vol}_M]\|_\infty}.$$

Proof The equality follows immediately from Propositions 12.6 and 12.7. □

12.3 Continuous Cohomology

It is apparent from Corollary 12.8 that proving the proportionality principle reduces to showing that the norm $\|[\mathrm{Vol}_M]\|_\infty$ only depends on the universal covering $\widetilde{M}$. The main ingredient of the proof (which we will carry out in Section 12.4) is the isomorphism between the Γ-invariant and the G-invariant cohomology groups of $\widetilde{M}$. This isomorphism, however, only holds in the setting of continuous cohomology, which we briefly introduce here.

12.3.1 Definition

If X is a topological space, denote by $S_k(X)$ the set of singular k-simplices $\Delta^k \to X$, endowed with the compact-open topology.

Definition 12.9 A cochain $\varphi \in C^k(X)$ is *continuous* if the restriction

$$\varphi|_{S_k(X)} \colon S_k(X) \longrightarrow \mathbb{R}$$

is continuous.

It is easy to see that the coboundary map sends continuous cochains to continuous cochains. We denote by $C_c^\bullet(X)$ the subcomplex of continuous cochains, and by $H_c^\bullet(X)$ the cohomology groups of this subcomplex.

12.3.2 Continuous Cochains as Injective Resolution of $\mathbb{R}$

Although the final theorem of this section will be crucial in our argument for the proportionality principle, we have decided to omit the proofs of the other intermediate results, since the insight they provide does not justify the effort required to report them. We will therefore simply state these results, referring the interested reader to Frigerio [2017].

Lemma 12.10 [Frigerio, 2017, proposition 8.4] *For every $k \geq 0$, the $\mathbb{R}\Gamma$-modules $C^k(\widetilde{M})$ and $C_c^k(\widetilde{M})$ are injective as $\mathbb{R}\Gamma$-modules.*

Proposition 12.11 [Frigerio, 2017, proposition 8.5] *The complexes*

$$0 \longrightarrow \mathbb{R} \longrightarrow C^0(\widetilde{M}) \longrightarrow C^1(\widetilde{M}) \longrightarrow \cdots \longrightarrow C^k(\widetilde{M}) \longrightarrow \cdots$$

and

$$0 \longrightarrow \mathbb{R} \longrightarrow C^0_c(\widetilde{M}) \longrightarrow C^1_c(\widetilde{M}) \longrightarrow \cdots \longrightarrow C^k_c(\widetilde{M}) \longrightarrow \cdots$$

are injective resolutions of the trivial $\mathbb{R}\Gamma$-module $\mathbb{R}$.

Proposition 12.12 [Frigerio, 2017, theorem 4.14] *Let A, B be $\mathbb{R}\Gamma$-modules, and let $I^\bullet$, $J^\bullet$ be injective resolutions of A, B, respectively. Let $f: A \to B$ be a morphism of $\mathbb{R}\Gamma$-modules. Then there exists a chain map $f^\bullet: I^\bullet \to J^\bullet$ that extends f, that is, makes the diagram*

$$
\begin{array}{ccccccccc}
0 & \longrightarrow & A & \longrightarrow & I^0 & \longrightarrow & I^1 & \longrightarrow & \cdots & \longrightarrow & I^k & \longrightarrow & \cdots \\
 & & \downarrow{\scriptstyle f} & & \downarrow{\scriptstyle f^0} & & \downarrow{\scriptstyle f^1} & & & & \downarrow{\scriptstyle f^k} & & \\
0 & \longrightarrow & B & \longrightarrow & J^0 & \longrightarrow & J^1 & \longrightarrow & \cdots & \longrightarrow & J^k & \longrightarrow & \cdots
\end{array}
$$

commute. Moreover, $f^\bullet$ is unique up to Γ-homotopy.

12.3.3 Continuous Cohomology versus Singular Cohomology

Denote by $C^\bullet(\widetilde{M})^\Gamma$ the subcomplex of $C^\bullet(\widetilde{M})$ given by Γ-invariant cochains, and by $H^\bullet(\widetilde{M})^\Gamma$ its cohomology groups. Similarly, denote by $C^\bullet_c(\widetilde{M})^\Gamma$ the subcomplex of continuous Γ-invariant cochains, and by $H^\bullet_c(\widetilde{M})^\Gamma$ its cohomology groups. We have the following:

Lemma 12.13 [Frigerio, 2017, lemma 8.2] *The covering map $p: \widetilde{M} \to M$ induces chain maps*

$$p^\bullet: C^\bullet(M) \longrightarrow C^\bullet(\widetilde{M})^\Gamma, \qquad p^\bullet: C^\bullet_c(M) \longrightarrow C^\bullet_c(\widetilde{M})^\Gamma,$$

which in turn induce isometric isomorphisms

$$H^\bullet(p^\bullet): H^\bullet(M) \longrightarrow H^\bullet(\widetilde{M})^\Gamma, \qquad H^\bullet(p^\bullet): H^\bullet_c(M) \longrightarrow H^\bullet_c(\widetilde{M})^\Gamma$$

in cohomology.

Proof It is straightforward to check that $p^\bullet: C^\bullet(M) \to C^\bullet(\widetilde{M})^\Gamma$ is an isometric isomorphism of cochain complexes. The only thing we are left to prove is that a cochain $\varphi \in C^k(M)$ is continuous if and only if $p^\bullet(\varphi)$ is continuous. In Frigerio [2011, lemma A.4] it is shown that $p_\bullet: S_k(\widetilde{M}) \to S_k(M)$ is a covering map, from which the claim readily follows. $\qquad\square$

Theorem 12.14 [Frigerio, 2017, proposition 8.7] *The inclusion of complexes* $i\colon \mathrm{C}^\bullet_c(M) \to \mathrm{C}^\bullet(M)$ *induces isometric isomorphisms*

$$\mathrm{H}^\bullet(i^\bullet)\colon \mathrm{H}^\bullet_c(M) \longrightarrow \mathrm{H}^\bullet(M)$$

in cohomology.

Proof Consider the following diagram:

$$
\begin{array}{ccccccccc}
0 & \longrightarrow & \mathbb{R} & \longrightarrow & \mathrm{C}^0_c(\widetilde{M}) & \longrightarrow & \mathrm{C}^1_c(\widetilde{M}) & \longrightarrow \cdots \longrightarrow & \mathrm{C}^k_c(\widetilde{M}) \longrightarrow \cdots \\
& & \downarrow{\scriptstyle \mathrm{id}_\mathbb{R}} & & \downarrow{\scriptstyle j^0} & & \downarrow{\scriptstyle j^1} & & \downarrow{\scriptstyle j^k} \\
0 & \longrightarrow & \mathbb{R} & \longrightarrow & \mathrm{C}^0(\widetilde{M}) & \longrightarrow & \mathrm{C}^1(\widetilde{M}) & \longrightarrow \cdots \longrightarrow & \mathrm{C}^k(\widetilde{M}) \longrightarrow \cdots
\end{array}
$$

where $j^\bullet\colon \mathrm{C}^\bullet_c(\widetilde{M}) \to \mathrm{C}^\bullet(\widetilde{M})$ is the inclusion map. By Proposition 12.11, the two rows are injective resolutions of the trivial $\mathbb{R}\Gamma$-module, and $j^\bullet$ is a norm non-increasing chain map extending the identity of $\mathbb{R}$. It can be shown that there exists a norm non-increasing chain map $\theta^\bullet\colon \mathrm{C}^\bullet(\widetilde{M}) \to \mathrm{C}^\bullet_c(\widetilde{M})$ extending the identity of $\mathbb{R}$. (See Frigerio [2017, theorem 4.16] for a detailed proof of this fact.) By the uniqueness property stated in Proposition 12.12, both $j^\bullet \circ \theta^\bullet\colon \mathrm{C}^\bullet(\widetilde{M}) \to \mathrm{C}^\bullet(\widetilde{M})$ and $\theta^\bullet \circ j^\bullet\colon \mathrm{C}^\bullet_c(\widetilde{M}) \to \mathrm{C}^\bullet_c(\widetilde{M})$ are Γ-homotopic to the identity. As a consequence, we get norm non-increasing maps in cohomology

$$\mathrm{H}^\bullet(j^\bullet)\colon \mathrm{H}^\bullet_c(\widetilde{M})^\Gamma \longrightarrow \mathrm{H}^\bullet(\widetilde{M})^\Gamma, \qquad \mathrm{H}^\bullet(\theta^\bullet)\colon \mathrm{H}^\bullet(\widetilde{M})^\Gamma \longrightarrow \mathrm{H}^\bullet_c(\widetilde{M})^\Gamma,$$

which are inverses of each other. Specifically, $\mathrm{H}^\bullet(j^\bullet)$ are isometric isomorphisms. Thanks to the commutative diagram

$$
\begin{array}{ccc}
\mathrm{C}^\bullet_c(M) & \xrightarrow{\ \ p^\bullet\ \ }_{\cong} & \mathrm{C}^\bullet_c(\widetilde{M}) \\
\downarrow{\scriptstyle i^\bullet} & & \downarrow{\scriptstyle j^\bullet} \\
\mathrm{C}^\bullet(M) & \xrightarrow{\ \ p^\bullet\ \ }_{\cong} & \mathrm{C}^\bullet(\widetilde{M}),
\end{array}
$$

whose horizontal arrows are isometric isomorphisms due to Lemma 12.13, we conclude that $\mathrm{H}^\bullet(i^\bullet)\colon \mathrm{H}^\bullet_c(M) \to \mathrm{H}^\bullet(M)$ are isometric isomorphisms as well. $\qquad \square$

12.4 The Proportionality Principle

12.4.1 The Transfer Map

Denote by $\mathrm{C}^\bullet_c(\widetilde{M})^\Gamma$, $\mathrm{C}^\bullet_c(\widetilde{M})^G$ the subcomplexes of $\mathrm{C}^\bullet_c(\widetilde{M})$ given respectively by Γ-invariant and G-invariant continuous cochains, and let $\mathrm{H}^\bullet_c(\widetilde{M})^\Gamma$, $\mathrm{H}^\bullet_c(\widetilde{M})^G$ be their respective cohomology groups. The inclusion $\mathrm{C}^\bullet_c(\widetilde{M})^G \to \mathrm{C}^\bullet_c(\widetilde{M})^\Gamma$ induces maps in cohomology

$$\mathrm{res}^\bullet \colon \ \mathrm{H}_c^\bullet(\widetilde{M})^G \longrightarrow \mathrm{H}_c^\bullet(\widetilde{M})^\Gamma$$

called *restriction maps*. Notice that these maps are norm non-increasing, if we endow $\mathrm{H}_c^\bullet(\widetilde{M})^G$ and $\mathrm{H}_c^\bullet(\widetilde{M})^\Gamma$ with the seminorms induced by those of $\mathrm{C}_c^\bullet(\widetilde{M})^G$ and $\mathrm{C}_c^\bullet(\widetilde{M})^\Gamma$. Our efforts will now be devoted to finding norm non-increasing left inverses of $\mathrm{res}^\bullet$, in order to show that the restriction maps are isometric embeddings.

By the Myers–Steenrod theorem [Myers and Steenrod, 1939], G admits a Lie group structure compatible with the compact-open topology. Therefore, there exists a left-invariant Borel measure μ on G, unique up to scaling (the *Haar measure*); since G contains a cocompact subgroup, the Haar measure is also right-invariant [Sauer, 2002, lemma 2.32]. Moreover, since Γ is a discrete subgroup of G and $M = \widetilde{M}/\Gamma$ is compact, there exists a relatively compact measurable subset $F \subseteq G$ such that $\{\gamma(F)\}_{\gamma \in \Gamma}$ is a locally finite partition of G; we will normalize the Haar measure in such a way that $\mu(F) = 1$.

We now proceed to define the *transfer map*. Given a continuous cochain $\varphi \in \mathrm{C}_c^k(\widetilde{M})$, set

$$\mathrm{trans}^k(\varphi)(s) = \int_F \varphi(g \cdot s)\,d\mu(g)$$

for every k-simplex s. Notice that this definition makes sense, since $\varphi(- \cdot s)$ is a continuous function from G to $\mathbb{R}$, and F is a relatively compact subset of G. Extending by linearity, we get maps

$$\mathrm{trans}^k \colon \ \mathrm{C}_c^k(\widetilde{M}) \longrightarrow \mathrm{C}^k(\widetilde{M}).$$

Proposition 12.15 [Frigerio, 2017, proposition 8.8] *The maps* trans^k *enjoy the following properties:*

1. *$\delta \circ \mathrm{trans}^k = \mathrm{trans}^{k+1} \circ \delta$, that is, $\mathrm{trans}^\bullet$ is a chain map (δ denotes the coboundary map);*
2. *for every $\varphi \in \mathrm{C}_c^k(\widetilde{M})$, the cochain $\mathrm{trans}^k(\varphi)$ is continuous;*
3. *if φ is Γ-invariant, then $\mathrm{trans}^k(\varphi)$ is G-invariant;*
4. *if φ is G-invariant, then $\mathrm{trans}^k(\varphi) = \varphi$.*

Proof Properties *1* and *4* are very easy to check. Let us consider the remaining ones:

2. Since $\widetilde{M}$ is a metric space and Δ^k is compact, the compact-open topology on $S_k(\widetilde{M})$ is induced by the distance

$$\mathrm{dist}(s, s') = \sup \Big\{ \mathrm{dist}_{\widetilde{M}}(s(x), s'(x)) : x \in \Delta^k \Big\}.$$

Let $s_0 \in S_k(\widetilde{M})$, and fix $\varepsilon > 0$. Since $\overline{F}$ is compact in G, the set $\overline{F} \cdot s_0 = \{g \cdot s_0 : g \in F\}$ is compact in $S_k(\widetilde{M})$. By continuity of φ, there exists $\eta > 0$ such that $|\varphi(s') - \varphi(s)| < \varepsilon$ for every $s \in \overline{F} \cdot s_0$, $s' \in S_k(\widetilde{M})$ with $\mathrm{dist}(s, s') < \eta$.

Let $s \in S_k(\widetilde{M})$ with $\mathrm{dist}(s, s_0) < \eta$. Since G is a group of isometries of $\widetilde{M}$, we have that $\mathrm{dist}(g \cdot s, g \cdot s_0) < \eta$ for every $g \in G$. But then

$$\left| \mathrm{trans}^k(\varphi)(s) - \mathrm{trans}^k(\varphi)(s_0) \right| \leq \int_F |\varphi(g \cdot s) - \varphi(g \cdot s_0)| \, d\mu(g) < \varepsilon;$$

hence the continuity of $\mathrm{trans}^k(\varphi)$.

3. Fix $s \in S_k(\widetilde{M})$, $g_0 \in G$. Since F is relatively compact in G, the same holds for $F \cdot g_0$ and $F \cdot g_0^{-1}$. We can therefore find elements $\gamma_1, \dots, \gamma_r \in \Gamma$ such that

$$F \cdot g_0 \subseteq \bigsqcup_{i=1}^{r} \gamma_i \cdot F \qquad \text{and} \qquad F \cdot g^{-1} \subseteq \bigsqcup_{i=1}^{r} \gamma_i^{-1} \cdot F.$$

Set $F_i = F \cap (\gamma_i^{-1} \cdot F \cdot g_0)$. It is immediate to check that

$$F = \bigsqcup_{i=1}^{r} F_i \qquad \text{and} \qquad F \cdot g_0 = \bigsqcup_{i=1}^{r} \gamma_i \cdot F_i.$$

By left and right G-invariance of μ and by Γ-invariance of φ, we get

$$\begin{aligned}
\mathrm{trans}^k(\varphi)(g_0 \cdot s) &= \int_F \varphi(gg_0 \cdot s) d\mu(g) \\
&= \int_{F \cdot g_0} \varphi(g \cdot s) d\mu(g) \\
&= \sum_{i=1}^{r} \int_{\gamma_i \cdot F_i} \varphi(g \cdot s) d\mu(g) \\
&= \sum_{i=1}^{r} \int_{F_i} \varphi(g \cdot s) d\mu(g) \\
&= \int_F \varphi(g \cdot s) \mu(g) = \mathrm{trans}^k(\varphi)(s).
\end{aligned} \qquad \square$$

Corollary 12.16 [Frigerio, 2017, proposition 8.9] *The restriction map*

$$\mathrm{res}^{\bullet} \colon \mathrm{H}_c^{\bullet}(\widetilde{M})^G \longrightarrow \mathrm{H}_c^{\bullet}(\widetilde{M})^{\Gamma}$$

is an isometric embedding.

Proof From Proposition 12.15, it readily follows that

$$\mathrm{trans}^{\bullet} \colon \mathrm{C}_c^{\bullet}(\widetilde{M})^{\Gamma} \longrightarrow \mathrm{C}_c^{\bullet}(\widetilde{M})^G$$

is a well-defined chain map that restricts to the identity on G-invariant cochains. It's also clear from the definition that $\|\mathrm{trans}^k(\varphi)\|_\infty \leq \|\varphi\|_\infty$ for every $\varphi \in C_c^k(\widetilde{M})^\Gamma$ (recall that $\mu(F) = 1$). Therefore, looking at the induced maps in cohomology, we get that

$$\mathrm{H}^\bullet(\mathrm{trans}^\bullet)\colon \mathrm{H}_c^\bullet(\widetilde{M})^\Gamma \longrightarrow \mathrm{H}_c^\bullet(\widetilde{M})^G$$

are norm non-increasing maps such that $\mathrm{H}^\bullet(\mathrm{trans}^\bullet)\circ\mathrm{res}^\bullet = \mathrm{id}$, thus completing the proof.   □

12.4.2 Proof of the Proportionality Principle

In light of Corollary 12.7, proving the proportionality principle is equivalent to showing that the seminorm of the volume coclass of M only depends on the isometry type of $\widetilde{M}$. Let Vol_M, $\mathrm{Vol}_{\widetilde{M}}$ be the volume cocycles of M, $\widetilde{M}$ respectively. We now list a few properties of these cocycles.

1. Vol_M is a continuous cochain. In fact, if $\{s_i\}_i$ is a sequence of n-simplices converging to s in the compact-open topology, then $\{\mathrm{str}_n(s_i)\}_i$ converges to $\mathrm{str}_n(s)$ in the C^1 topology. The claim follows from the fact that integration is continuous with respect to the C^1 topology. Obviously, the same holds for $\mathrm{Vol}_{\widetilde{M}}$.
2. Since $p\colon \widetilde{M} \to M$ is a local isometry, $p^\bullet(\mathrm{Vol}_M) = \mathrm{Vol}_{\widetilde{M}}$.
3. The cochain $\mathrm{Vol}_{\widetilde{M}}$ is G-invariant and, even more so, Γ-invariant.

Denote by $[\mathrm{Vol}_M]_c$ the coclass of Vol_M in the continuous cohomology group $\mathrm{H}_c^n(M)$, and by $[\mathrm{Vol}_{\widetilde{M}}]_c^\Gamma$, $[\mathrm{Vol}_{\widetilde{M}}]_c^G$ the coclasses of $\mathrm{Vol}_{\widetilde{M}}$ in the cohomology groups $\mathrm{H}_c^n(\widetilde{M})^\Gamma$, $\mathrm{H}_c^n(\widetilde{M})^G$, respectively. The following theorem will complete our proof of Gromov's proportionality principle, since the ℓ^∞-seminorm of the coclass $[\mathrm{Vol}_{\widetilde{M}}]_c^G$ only depends on the isometry type of $\widetilde{M}$.

Theorem 12.17 *We have* $\|[\mathrm{Vol}_M]\|_\infty = \|[\mathrm{Vol}_{\widetilde{M}}]_c^G\|_\infty.$

Proof Notice that all the maps in the following diagram are either isometric isomorphisms or isometric embeddings, due to Theorem 12.14, Lemma 12.13, and Corollary 12.16, respectively:

$$\mathrm{H}^n(M) \xleftarrow{\ \mathrm{H}^n(i^\bullet)\ } \mathrm{H}_c^n(M) \xrightarrow{\ \mathrm{H}^n(p^\bullet)\ } \mathrm{H}_c^n(\widetilde{M})^\Gamma \xleftarrow{\ \mathrm{res}^n\ } \mathrm{H}_c^n(\widetilde{M})^G$$

Moreover, $[\mathrm{Vol}_M]$ on the left corresponds to $[\mathrm{Vol}_{\widetilde{M}}]_c^G$ on the right; hence the equality in the statement.   □

12.5 Simplicial Volume of Hyperbolic Manifolds

Let M be a closed hyperbolic n-manifold. Corollary 12.7 implies that the ratio $\|M\| / \mathrm{vol}(M)$ is a constant depending only on n; by Theorem 12.17, this constant can be computed explicitly as $\left\| [\mathrm{Vol}_{\mathbb{H}^n}]_c^G \right\|_\infty$. In this section, our efforts will be devoted to finding the value of this seminorm.

12.5.1 Simplices in $\mathbb{H}^n$

For the convenience of the reader, we recall here a few notions from Chapter 2 (compare with Remark 2.2 and Section 2.1). We start with some elementary properties of simplices in hyperbolic space. Denote by $\overline{\mathbb{H}}^n = \mathbb{H}^n \cup \partial \mathbb{H}^n$ the usual compactification of hyperbolic space. A *geodesic k-simplex* in $\overline{\mathbb{H}}^n$ is the convex hull of some $(k + 1)$ points, called *vertices* of the simplex. A geodesic simplex is *finite* if all its vertices lie in $\mathbb{H}^n$, *ideal* if all its vertices lie in $\partial \mathbb{H}^n$, and *regular* if every permutation of its vertices is induced by an isometry of $\mathbb{H}^n$.

It is easy to see that geodesic simplices are exactly the images of straight simplices. Moreover, for every positive real number ℓ there is exactly one (up to isometry) finite regular geodesic n-simplex of edgelength ℓ, denoted by τ_ℓ. Finally, there is exactly one (always up to isometry) ideal regular geodesic n-simplex, which we will denote by τ_∞.

The following (hard) theorem gives a characterization of geodesic simplices of maximal volume (see also Thurston [1979, chapter 7] for a proof in dimension 3).

Theorem 12.18 [Haagerup and Munkholm, 1981] *Denote by $v_n = \mathrm{vol}(\tau_\infty)$ the volume of the ideal regular geodesic n-simplex in $\mathbb{H}^n$ (see Section 2.1). Let Δ be any geodesic n-simplex. Then $\mathrm{vol}(\Delta) \leq v_n$, and the inequality is strict unless Δ is ideal and regular.*

Since the volume of a geodesic simplex is continuous with respect to the position of its vertices, we have that

$$\lim_{\ell \to \infty} \mathrm{vol}(\tau_\ell) = \mathrm{vol}(\tau_\infty) = v_n.$$

12.5.2 Alternating Cochains

We now make a short digression to introduce the concept of alternating cochain, which will be useful in the proof of the main result of this section. This has already been mentioned in the context of (bounded) cohomology of groups both in the introduction (Remark 13) and in Chapter 9.

Definition 12.19 Let X be a topological space. A cochain $\varphi \in C^k(X)$ is *alternating* if, given any permutation $\sigma \in S_{k+1}$, the equality

$$\varphi(s \circ \overline{\sigma}) = \text{sign}(\sigma) \cdot \varphi(s)$$

holds, where $\overline{\sigma} \colon \Delta^k \to \Delta^k$ is the affine map inducing the permutation σ on the vertices of the standard simplex Δ^k and $s \colon \Delta^k \to X$ is a singular simplex.

It is easy to see that the coboundary map sends alternating cochains to alternating cochains. We denote by $C^{\bullet}_{alt}(X)$ the subcomplex of alternating cochains. We can define a chain "projection" $\text{alt}^{\bullet} \colon C^{\bullet}(X) \to C^{\bullet}_{alt}(X)$ by setting

$$\text{alt}^k(\varphi)(s) = \frac{1}{(k+1)!} \sum_{\sigma \in S_{k+1}} \text{sign}(\sigma) \cdot \varphi(s \circ \overline{\sigma}).$$

This projection is norm non-increasing and restricts to the identity on alternating cochains.

12.5.3 Seminorm of the Volume Coclass

We are now ready to compute the proportionality constant between simplicial and Riemannian volume for hyperbolic manifolds.

Proposition 12.20 [Thurston, 1979; Gromov, 1982] *Let M be a closed hyperbolic manifold. Then*

$$\frac{\|M\|}{\text{vol}(M)} = \frac{1}{v_n}.$$

Proof Thanks to Theorem 12.17, it is enough to show that $\left\| [\text{Vol}_{\mathbb{H}^n}]^G_c \right\|_{\infty} = v_n$.

($\leq$) For this inequality, simply notice that

$$\left\| [\text{Vol}_{\mathbb{H}^n}]^G_c \right\|_{\infty} \leq \|\text{Vol}_{\mathbb{H}^n}\|_{\infty} \leq v_n,$$

since by Theorem 12.18 every straight simplex has volume at most v_n.

($\geq$) For the other inequality, some more work is required [Frigerio, 2017, section 8.11]. By definition, we have

$$\left\| [\mathrm{Vol}_{\mathbb{H}^n}]_c^G \right\|_\infty = \inf \left\{ \left\| \mathrm{Vol}_{\mathbb{H}^n} + \delta \varphi \right\|_\infty : \varphi \in C_c^{n-1}(\mathbb{H}^n)^G \right\},$$

where δ is the coboundary map. Observe that the volume cochain is alternating and that the chain projection $\mathrm{alt}^\bullet$ preserves continuity and G-invariance of cochains. Therefore

$$\left\| [\mathrm{Vol}_{\mathbb{H}^n}]_c^G \right\|_\infty \geq \inf \left\{ \left\| \mathrm{alt}^n(\mathrm{Vol}_{\mathbb{H}^n} + \delta \varphi) \right\|_\infty : \varphi \in C_c^{n-1}(\mathbb{H}^n)^G \right\}$$

$$= \inf \left\{ \left\| \mathrm{Vol}_{\mathbb{H}^n} + \delta \, \mathrm{alt}^{n-1}(\varphi) \right\|_\infty : \varphi \in C_c^{n-1}(\mathbb{H}^n)^G \right\}$$

$$= \inf \left\{ \left\| \mathrm{Vol}_{\mathbb{H}^n} + \delta \psi \right\|_\infty : \psi \in C_{c,alt}^{n-1}(\mathbb{H}^n)^G \right\}.$$

In other words, we can restrict our attention to the boundaries of continuous G-invariant alternating $(n-1)$-cochains. Let ψ be such a cochain, and consider a regular finite geodesic n-simplex τ_ℓ. Denote by $s_\ell \colon \Delta^n \to \mathbb{H}^n$ its barycentric parametrization[3], and let $\partial_i s_\ell$ be the ith face of s_ℓ. Fix an odd permutation $\sigma \in S_n$; since τ_ℓ is regular, there exists an isometry g of $\mathbb{H}^n$ inducing the permutation σ on the vertices of $\partial_i s_\ell$. Up to composing with the reflection about the hyperplane containing $\partial_i s_\ell$, we can assume that g is orientation-preserving. Since $\partial_i s_\ell$ is a barycentric parametrization, it is easy to see that $g \circ \partial_i s_\ell = \partial_i s_\ell \circ \overline{\sigma}$ (isometries preserve convex combinations). Exploiting the fact that ψ is both alternating and G-invariant, we get that

$$\psi(\partial_i s_\ell) = \psi(g \circ \partial_i s_\ell) = \psi(\partial_i s_\ell \circ \overline{\sigma}) = -\psi(\partial_i s_\ell),$$

from which $\psi(\partial_i s_\ell) = 0$ and therefore $\psi(\partial_{s_\ell}) = 0$. We then conclude that

$$\left\| [\mathrm{Vol}_{\mathbb{H}^n}]_c^G \right\|_\infty \geq \inf \left\{ \left\| \mathrm{Vol}_{\mathbb{H}^n} + \delta \psi \right\|_\infty : \psi \in C_{c,alt}^{n-1}(\mathbb{H}^n)^G \right\}$$

$$\geq \inf \left\{ \left| (\mathrm{Vol}_{\mathbb{H}^n} + \delta \psi)(s_\ell) \right| : \psi \in C_{c,alt}^{n-1}(\mathbb{H}^n)^G \right\}$$

$$= \left| \mathrm{Vol}_{\mathbb{H}^n}(s_\ell) \right| = \mathrm{vol}(\tau_\ell).$$

Taking the limit $\ell \to \infty$, we finally get $\left\| [\mathrm{Vol}_{\mathbb{H}^n}]_c^G \right\|_\infty \geq v_n$. $\qquad \square$

---

[3] To be more explicit, our s_ℓ is exactly the *straightening* of τ_ℓ as defined in Section 2.3.

References

Abert, M., Bergeron, N., Fraczyk, M., and Gaboriau, D. 2021. On homology torsion growth. *arXiv preprint arXiv:2106.13051*.

Agol, I. 2013. The virtual Haken conjecture. *Doc. Math.*, **18**, 1045–1087. With an appendix by Agol, D. Groves, and J. Manning.

Albuquerque, P. 1999. Patterson-Sullivan theory in higher rank symmetric spaces. *Geom. Funct. Anal.*, **9**(1), 1–28.

Alonso, J. M., Brady, T., Cooper, D. et al. 1991. Notes on word hyperbolic groups. Pages 3–63 in *Group theory from a geometrical viewpoint (Trieste, 1990)*. World Sci. Publ., River Edge, NJ. Edited by Short.

Arveson, W. 1976. *An invitation to C*-algebras*. Springer-Verlag, New York-Heidelberg. Graduate Texts in Mathematics, No. 39.

Babenko, I. K. 1992. Asymptotic invariants of smooth manifolds. *Izv. Ross. Akad. Nauk Ser. Mat.*, **56**(4), 707–751.

Balacheff, F., and Karam, S. 2019. Macroscopic Schoen conjecture for manifolds with nonzero simplicial volume. *Trans. Amer. Math. Soc.*, **372**(10), 7071–7086.

Ballmann, W. 1985. Nonpositively curved manifolds of higher rank. *Ann. of Math. (2)*, **122**(3), 597–609.

Bavard, C. 1991. Longueur stable des commutateurs. *Enseign. Math. (2)*, **37**(1–2), 109–150.

Benedetti, R., and Petronio, C. 1992. *Lectures on hyperbolic geometry*. Universitext. Springer-Verlag, Berlin.

Berger, M. 2003. *A panoramic view of Riemannian geometry*. Springer-Verlag, Berlin.

Bergeron, N., and Venkatesh, A. 2013. The asymptotic growth of torsion homology for arithmetic groups. *J. Inst. Math. Jussieu*, **12**(2), 391–447.

Besson, G., Courtois, G., and Gallot, S. 1995. Entropies et rigidités des espaces localement symétriques de courbure strictement négative. *Geom. Funct. Anal.*, **5**(5), 731–799.

Bestvina, M., and Fujiwara, K. 2002. Bounded cohomology of subgroups of mapping class groups. *Geom. Topol.*, **6**(1), 69–89.

Bestvina, M., and Fujiwara, K. 2009. A characterization of higher rank symmetric spaces via bounded cohomology. *Geom. Funct. Anal.*, **19**(1), 11–40.

Bestvina, M., Bromberg, K., and Fujiwara, K. 2016. Stable commutator length on mapping class groups. *Ann. Inst. Fourier (Grenoble)*, **66**(3), 871–898.

Bloch, S. J. 2000. *Higher regulators, algebraic K-theory, and zeta functions of elliptic curves*. CRM Monograph Series, vol. 11. American Mathematical Society, Providence, RI.

Braun, S. 2018. *Simplicial volume and macroscopic scalar curvature*. PhD thesis, Karlsruher Institut für Technologie.

Bridson, M. R. 2006. Non-positive curvature and complexity for finitely presented groups. Pages 961–987 in *International Congress of Mathematicians*, vol. 2. European Mathematical Society, Madrid.

Brooks, R. 1981. Some remarks on bounded cohomology. Pages 53–63 in *Riemann surfaces and related topics: Proceedings of the 1978 Stony Brook Conference (State Univ. New York, Stony Brook, NY, 1978)*, vol. 97. Princeton University Press Annals of Mathematics Studies, Princeton.

Brunnbauer, M. 2008. Homological invariance for asymptotic invariants and systolic inequalities. *Geom. Funct. Anal.*, **18**(4), 1087–1117.

Bucher, M. 2007. Simplicial volume of locally symmetric spaces covered by $SL_3\mathbb{R}/SO(3)$. *Geom. Dedicata*, **125**, 203–224.

Bucher, M. 2008. The simplicial volume of closed manifolds covered by $\mathbb{H}^2 \times \mathbb{H}^2$. *J. Topol.*, **1**(3), 584–602.

Bucher, M. 2009. Simplicial volume of products and fiber bundles. Pages 79–86 in *Discrete groups and geometric structures*. Contemp. Math., vol. 501. American Mathematical Society, Providence, RI.

Bucher, M., Burger, M., and Iozzi, A. 2013. A dual interpretation of the Gromov–Thurston proof of Mostow rigidity and volume rigidity for representations of hyperbolic lattices. Pages 47–76 in *Trends in harmonic analysis*. Springer INdAM Ser. Springer, Milan.

Bucher, M., Frigerio, R., and Hartnick, T. 2016. A note on semi-conjugacy for circle actions. *Enseign. Math.*, **62**(3-4), 317–360.

Bulteau, G. 2015. Cycles géométriques réguliers. *Bull. Soc. Math. France*, **143**(4), 727–761.

Burger, M., and Iozzi, A. 2002. Boundary maps in bounded cohomology. *Geom. Funct. Anal.*, **12**(2), 281–292. Appendix to "Continuous bounded cohomology and applications to rigidity theory" by Marc Burger and Nicolas Monod.

Burger, M., and Iozzi, A. 2004. Bounded Kähler class rigidity of actions on Hermitian symmetric spaces. *Ann. Sci. Éc. Norm. Sup.*, **37**(1), 77–103.

Burger, M., and Iozzi, A. 2009. A useful formula from bounded cohomology. Pages 243–292 in *Géométries à courbure négative ou nulle, groupes discrets et rigidités*. Sémin. Congr., vol. 18. Soc. Math. France, Paris.

Burger, M., and Monod, N. 1999. Bounded cohomology of lattices in higher rank Lie groups. *J. Eur. Math. Soc. (JEMS)*, **1**(2), 199–235.

Burger, M., and Monod, N. 2002. Continuous bounded cohomology and applications to rigidity theory. *Geom. Funct. Anal.*, **12**(2), 219–280.

Burger, M., and Mozes, S. 1996. CAT(-1)-spaces, divergence groups and their commensurators. *J. Amer. Math. Soc.*, **9**(1), 57–93.

Burger, M., and Mozes, S. 2000. Lattices in product of trees. *Inst. Hautes Études Sci. Publ. Math.*, 151–194 (2001).

Burger, M., Iozzi, A., and Wienhard, A. 2010. Surface group representations with maximal Toledo invariant. *Ann. of Math. (2)*, **172**, 517–566.

Burns, K., and Spatzier, R. 1987. Manifolds of nonpositive curvature and their buildings. *Inst. Hautes Études Sci. Publ. Math.*, **65**, 35–59.

Calegari, D. 2007. Stable commutator length in subgroups of $PL^+(I)$. *Pacific J. Math.*, **232**(2), 257–262.

Calegari, D. 2008. What is ... stable commutator length? *Notices Amer. Math. Soc.*, **55**(9), 1100–1101.

Calegari, D. 2009a. *scl*. MSJ Memoirs, vol. 20. Mathematical Society of Japan, Tokyo.

Calegari, D. 2009b. Stable commutator length is rational in free groups. *J. Amer. Math. Soc.*, **22**(4), 941–961.

Calegari, D., and Fujiwara, K. 2010. Stable commutator length in word-hyperbolic groups. *Groups Geom. Dyn.*, **4**(1), 59–90.

Calegari, D., and Walker, A. 2009. scallop. *Computer program.*

Campagnolo, C., and Corro, D. 2021. Integral foliated simplicial volume and circle foliations. *https://doi.org/10.1142/S1793525321500266*, 1–30. online ready in *J. Topol. Anal.*

Cashen, C. H., and Hoffmann, C. 2020. Short, highly imprimitive words yield hyperbolic one-relator groups. *Exp. Math., to appear. arXiv:2006.15923.*

Christopher H. Cashen and Charlotte Hoffmann, 2021. Short, Highly Imprimitive Words Yield Hyperbolic One-Relator Groups. *Experimental Mathematics*, 1–10 Taylor & Francis.

Chatterji, I., Fernós, T., and Iozzi, A. 2016. The median class and superrigidity of actions on CAT(0) cube complexes. *J. Topol.*, **9**(2), 349–400.

Cheeger, J., and Gromov, M. 1986. L_2-cohomology and group cohomology. *Topology*, **25**(2), 189–215.

Chen, L. 2018. Spectral gap of scl in free products. *Proc. Amer. Math. Soc.*, **146**(7), 3143–3151.

Chen, L., and Heuer, N. 2019. Spectral gap of scl in graphs of groups and 3-manifolds. *arXiv preprint arXiv:1910.14146.*

Chen, L., and Heuer, N. 2020. Stable commutator length in right-angled Artin and Coxeter groups. *arXiv preprint arXiv:2012.04088.*

Chen, L. 2020. scl in graphs of groups. *Invent. Math.*, **221**(2), 329–396.

Clay, M., Forester, M., and Louwsma, J. 2016. Stable commutator length in Baumslag-Solitar groups and quasimorphisms for tree actions. *Trans. Amer. Math. Soc.*, **368**(7), 4751–4785.

Clerc, J.-L., and Ørsted, B. 2003. The Gromov norm of the Kaehler class and the Maslov index. *Asian J. Math.*, **7**, 269–295.

Connell, C., and Farb, B. 2003. The degree theorem in higher rank. *J. Differential Geom.*, **65**(1), 19–59.

Connell, C., and Wang, S. 2019. Positivity of simplicial volume for nonpositively curved manifolds with a Ricci-type curvature condition. *Groups Geom. Dyn.*, **13**(3), 1007–1034.

Connell, C., and Wang, S. 2020. Some remarks on the simplicial volume of nonpositively curved manifolds. *Math. Ann.*, **377**(3-4), 969–987.

Culler, M. 1981. Using surfaces to solve equations in free groups. *Topology*, **20**(2), 133–145.

Dahmani, F., Guirardel, V., and Osin, D. 2017. Hyperbolically embedded subgroups and rotating families in groups acting on hyperbolic spaces. *Mem. Amer. Math. Soc.*, **245**(1156), v+152.

De la Cruz Mengual, C. 2019. *On Bounded-Cohomological Stability for Classical Groups*. PhD thesis, ETH Zürich.

Dupont, J. L. 1979. Bounds for characteristic numbers of flat bundles. Pages 109–119 in *Algebraic Topology, Aarhus 1978 (Proc. Sympos., Univ. Aarhus, Aarhus, 1978)*. Lecture Notes in Math., vol. 763. Springer, Berlin.

Eilenberg, S., and Zilber, J. A. 1950. Semi-simplicial complexes and singular homology. *Ann. of Math.*, **51**, 499–513.

Epstein, D. B. A., and Fujiwara, K. 1997. The second bounded cohomology of word-hyperbolic groups. *Topology*, **36**(6), 1275–1289.

Farb, B., and Margalit, D. 2012. *A primer on mapping class groups*. Princeton Mathematical Series, vol. 49. Princeton University Press, Princeton, NJ.

Farb, B., and Masur, H. 1998. Superrigidity and mapping class groups. *Topology*, **37**(6), 1169–1176.

Fauser, D. 2021. Integral foliated simplicial volume and S^1-actions. *Forum Math.*, **33**(3), 773–788.

Fauser, D., Friedl, S., and Löh, C. 2019. Integral approximation of simplicial volume of graph manifolds. *Bull. Lond. Math. Soc.*, **51**(4), 715–731.

Fauser, D., Löh, C., Moraschini, M., and Quintanilha, J. P. 2021. Stable integral simplicial volume of 3-manifolds. *J. Topol.*, **14**(2), 608–640.

Fernós, T., Forester, M., and Tao, J. 2019. Effective quasimorphisms on right-angled Artin groups. *Ann. Inst. Fourier (Grenoble)*, **69**(4), 1575–1626.

Forester, M., and Malestein, J. 2020. *On stable commutator length of non-filling curves in surfaces*. Available at https://math.ou.edu/~forester/papers/surfaces16.pdf, last accessed June 26, 2022.

Forester, M., Soroko, I., and Tao, J. 2020. Genus bounds in right-angled Artin groups. *Publ. Mat.*, **64**(1), 233–253.

Fournier-Facio, F. and Lodha, Y. 2021. Second bounded cohomology of groups acting on 1-manifolds and applications to spectrum problems. *arXiv preprint arXiv:2111.07931*.

Fournier-Facio, F., Löh, C., and Moraschini, M. 2021. Bounded cohomology of finitely presented groups: vanishing, non-vanishing and computability. *arXiv preprint arXiv:2106.13567*.

Fournier-Facio, F., Löh, C., and Moraschini, M. 2022. Bounded cohomology and binate groups. *J. Aust. Math. Soc.* Cambridge University Press, 1–36.

Francaviglia, S., Frigerio, R., and Martelli, B. 2012. Stable complexity and simplicial volume of manifolds. *J. Topol.*, **5**(4), 977–1010.

Franceschini, F., Frigerio, R., Pozzetti, M. B., and Sisto, A. 2019. The zero norm subspace of bounded cohomology of acylindrically hyperbolic groups. *Comment. Math. Helv.*, **94**(1), 89–139.

Frigerio, R. 2011. (Bounded) continuous cohomology and Gromov's proportionality principle. *Manuscripta Math.*, **134**(3–4), 435–474.

Frigerio, R. 2017. *Bounded cohomology of discrete groups*. Mathematical Surveys and Monographs, vol. 227. American Mathematical Society, Providence, RI.

Frigerio, R., and Moraschini, M. 2019. Gromov's theory of multicomplexes with applications to bounded cohomology and simplicial volume. *Mem. Amer. Math. Soc.*, *to appear. arXiv:1808.07307.pdf.*

Frigerio, R., Pozzetti, M. B., and Sisto, A. 2015. Extending higher-dimensional quasicocycles. *J. Topol.*, **8**(4), 1123–1155.

Frigerio, R., Löh, C., Pagliantini, C., and Sauer, R. 2016. Integral foliated simplicial volume of aspherical manifolds. *Israel J. Math.*, **216**(2), 707–751.

Fritsch, R., and Piccinini, A. 1990. *Cellular structures in topology*. Cambridge Studies in Advanced Mathematics, no. 19. Cambridge University Press, New York.

Gaboriau, D. 2002. Invariants l^2 de relations d'équivalence et de groupes. *Publ. Math. Inst. Hautes Études Sci.*, 93–150.

Ghys, E. 2001. Groups acting on the circle. *Enseign. Math.*, **47**(3/4), 329–408.

Goerss, P. G., and Jardine, J. F. 1999. *Simplicial homotopy theory*. Progress in Mathematics. Birkhäuser Verlag, Basel.

Goncharov, A. B. 1993. Explicit construction of characteristic classes. Pages 169–210 in *I. M. Gelfand Seminar*. Adv. Soviet Math., vol. 16. American Mathematical Society, Providence, RI.

Grandis, M. 2001a. Finite sets and symmetric simplicial sets. *Theory Appl. Categ.*, **8**(8), 244–252.

Grandis, M. 2001b. Higher functors for simplicial sets. *Cahiers Topologie Géom. Différentielle Catég.*, **42**, 101–136.

Gromov, M. 1981. Hyperbolic manifolds (according to Thurston and Jørgensen). Pages 40–53 in *Bourbaki Seminar, Vol. 1979/80*. Lecture Notes in Math., vol. 842. Springer, Berlin-New York.

Gromov, M. 1982. Volume and bounded cohomology. *Inst. Hautes Études Sci. Publ. Math.*, 5–99 (1983).

Gromov, M. 1983. Filling Riemannian manifolds. *J. Differential Geom.*, **18**(1), 1–147.

Gromov, M. 1987. Hyperbolic groups. Pages 75–263 in *Essays in group theory*. Math. Sci. Res. Inst. Publ., vol. 8. Springer, New York.

Gromov, M. 1993. Asymptotic invariants of infinite groups. Pages 1–295 in *Geometric group theory, Vol. 2 (Sussex, 1991)*. London Math. Soc. Lecture Note Ser., vol. 182. Cambridge University Press, Cambridge.

Gromov, M. 1996. Systoles and intersystolic inequalities. Pages 291–362 in *Actes de la Table Ronde de Géométrie Différentielle (Luminy, 1992)*. Sémin. Congr., vol. 1. Soc. Math. France, Paris.

Gromov, M. 1999. *Metric structures for Riemannian and non-Riemannian spaces*. Progress in Mathematics, vol. 152. Birkhäuser Boston, Inc., Boston, MA. Based on the 1981 French original [MR0682063 (85e:53051)], with appendices by M. Katz, P. Pansu and S. Semmes, translated from French by Sean Michael Bates.

Guichardet, A. 1980. *Cohomologie des groupes topologiques et des algèbres de Lie*. Textes Mathématiques [Mathematical Texts], vol. 2. CEDIC, Paris.

Guth, L. 2010. Metaphors in systolic geometry. Pages 745–768 in *Proceedings of the International Congress of Mathematicians. Volume II*. Hindustan Book Agency, New Delhi.

Haagerup, U., and Munkholm, H. J. 1981. Simplices of maximal volume in hyperbolic n-space. *Acta Math.*, **147**(1-2), 1–11.

Hamenstädt, U. 2008. Bounded cohomology and isometry groups of hyperbolic spaces. *J. Eur. Math. Soc. (JEMS)*, **10**(2), 315–349.

Hartnick, T., and Ott, A. 2015. Bounded cohomology via partial differential equations, I. *Geom. Topol.*, **19**(6), 3603–3643.

Harvey, W. J. 1981. Boundary structure of the modular group. Pages 245–251 in *Riemann surfaces and related topics: Proceedings of the 1978 Stony Brook Conference (State Univ. New York, Stony Brook, N.Y., 1978)*. Ann. of Math. Stud., vol. 97. Princeton University Press, Princeton, NJ.

Hatcher, A. 2002. *Algebraic topology*. Cambridge University Press, Cambridge.

Helgason, S. 1978. *Differential geometry, Lie groups, and symmetric spaces*. Pure and Applied Mathematics, vol. 80. Academic Press, Inc. [Harcourt Brace Jovanovich, Publishers], New York and London.

Hempel, J. 1987. Residual finiteness for 3-manifolds. Pages 379–396 in *Combinatorial group theory and topology (Alta, Utah, 1984)*. Ann. of Math. Stud., vol. 111. Princeton University Press, Princeton, NJ.

Heuer, N. 2019a. *Constructions in stable commutator length and bounded cohomology*. PhD thesis, University of Oxford.

Heuer, N. 2019b. The full spectrum of scl on recursively presented groups. *Geom. Dedicata, to appear. arXiv:1909.01309*.

Heuer, N. 2019c. Gaps in scl for amalgamated free products and RAAGs. *Geom. Funct. Anal.*, **29**(1), 198–237.

Heuer, N., and Löh, C. 2020. Transcendental simplicial volumes. *Ann. Inst. Fourier (Grenoble), to appear. arXiv:1911.06386*.

Heuer, N., and Löh, C. 2021a. The spectrum of simplicial volume. *Invent. Math.*, **223**(1), 103–148.

Heuer, N., and Löh, C. 2021b. The spectrum of simplicial volume of non-compact manifolds. *Geom. Dedicata*, 1–11.

Huber, T. 2013. *Rotation quasimorphisms for surfaces*. PhD thesis, ETH Zürich.

Hull, M., and Osin, D. 2013. Induced quasicocycles on groups with hyperbolically embedded subgroups. *Algebr. Geom. Topol.*, **13**(5), 2635–2665.

Inoue, H., and Yano, K. 1982. The Gromov invariant of negatively curved manifolds. *Topology*, **21**(1), 83–89.

Iozzi, A. 2002. Bounded cohomology, boundary maps, and rigidity of representations into $\mathrm{Homeo}_+(\mathbf{S}^1)$ and $\mathrm{SU}(1, n)$. Pages 237–260 in *Rigidity in dynamics and geometry (Cambridge, 2000)*. Springer, Berlin.

Ivanov, N. V. 1987. Foundations of the theory of bounded cohomology. *J. Soviet Math.*, **37**(3), 1090–1115.

Ivanov, N. V. 2017. Notes on the bounded cohomology theory. *arXiv preprint arXiv:1708.05150*.

Johnson, B. E. 1972. *Cohomology in Banach algebras*. Memoirs of the American Mathematical Society, No. 127. American Mathematical Society, Providence, RI.

Kammeyer, H. 2019. *Introduction to ℓ^2-invariants*. Lecture Notes in Mathematics, vol. 2247. Springer, Cham.

Knapp, A. W. 2002. *Lie groups beyond an introduction*. 2nd ed. Progress in Mathematics, vol. 140. Birkhäuser Boston, Boston, MA.

Knieper, G. 1997. On the asymptotic geometry of nonpositively curved manifolds. *Geom. Funct. Anal.*, **7**(4), 755–782.

Lafont, J.-F., and Schmidt, B. 2006. Simplicial volume of closed locally symmetric spaces of non-compact type. *Acta Math.*, **197**(1), 129–143.

Lafont, J.-F., and Wang, S. 2019. Barycentric straightening and bounded cohomology. *J. Eur. Math. Soc. (JEMS)*, **21**(2), 381–403.

Löh, C. 2006a. Measure homology and singular homology are isometrically isomorphic. *Math. Z.*, **253**(1), 197.

Löh, C. 2006b. *Simplicial volume and L^2-Betti numbers*. Available at www.mathematik.uni-regensburg.de/loeh/seminars/ goettingen.ps, last accessed October 21, 2021.

Löh, C., and Pagliantini, C. 2016. Integral foliated simplicial volume of hyperbolic 3-manifolds. *Groups Geom. Dyn.*, **10**(3), 825–865.

Löh, C., and Sauer, R. 2009. Simplicial volume of Hilbert modular varieties. *Comment. Math. Helv.*, **84**, 457–470.

Löh, C., Moraschini, M., and Raptis, G. 2022. On the simplicial volume and the Euler characteristic of (aspherical) manifolds. *Res. Math. Sci.*, **9**, 44.

Loos, O. 1969a. *Symmetric spaces. I: General theory*. W. A. Benjamin, New York and Amsterdam.

Loos, O. 1969b. *Symmetric spaces. II: Compact spaces and classification*. W. A. Benjamin, New York and Amsterdam.

Louder, L., and Wilton, H. 2022. Negative immersions for one-relator groups. *Duke Mathematical Journal*, Duke University Press, **171**(3), 547–594.

Lück, W. 1994. Approximating L^2-invariants by their finite-dimensional analogues. *Geom. Funct. Anal.*, **4**(4), 455–481.

Lück, W. 2002. L^2-invariants: theory and applications to geometry and K-theory. Ergebnisse der Mathematik und ihrer Grenzgebiete. 3. Folge. A Series of Modern Surveys in Mathematics, vol. 44. Springer–Verlag, Berlin.

Lück, W. 2005. Survey on classifying spaces for families of subgroups. Pages 269–322 in *Infinite groups: geometric, combinatorial and dynamical aspects*. Progr. Math., vol. 248. Birkhäuser, Basel.

Malcev, A. 1940. On isomorphic matrix representations of infinite groups. *Rec. Math. [Mat. Sbornik] N.S.*, **8**(50), 405–422.

Martelli, B. 2016. An introduction to geometric topology. *arXiv preprint arXiv:1610.02592*.

Masur, H. A., and Minsky, Y. N. 1999. Geometry of the complex of curves. I. Hyperbolicity. *Invent. Math.*, **138**(1), 103–149.

Matsumoto, S. 1987. Some remarks on foliated S^1-bundles. *Invent. Math.*, **90**, 343–358.

Matsumoto, S., and Morita, S. 1985. Bounded cohomology of certain groups of homeomorphisms. *Proc. Amer. Math. Soc.*, **94**(3), 539–544.

May, J. P. 1992. *Simplicial objects in algebraic topology*. Chicago Lectures in Mathematics. University of Chicago Press, Chicago, IL.

Milnor, J. 1957. The geometric realization of a semi-simplicial complex. *Ann. of Math.*, **65**(2), 357–362.

Mimura, M., and Toda, H. 1991. *Topology of Lie groups. I, II*. Translations of Mathematical Monographs, vol. 91. American Mathematical Society, Providence, RI. Translated from the 1978 Japanese edition by the authors.

Mineyev, I. 2002. Bounded cohomology characterizes hyperbolic groups. *Q. J. Math.*, **53**(1), 59–73.

Mineyev, I., Monod, N., and Shalom, Y. 2004. Ideal bicombings for hyperbolic groups and applications. *Topology*, **43**(6), 1319–1344.

Mitsumatsu, Y. 1984. Bounded cohomology and ll-homology of surfaces. *Topology*, **23**(4), 465–471.

Monod, N. 2004. Stabilization for SL_n in bounded cohomology. Pages 191–202 in *Discrete geometric analysis*. Contemp. Math., vol. 347. American Mathematical Society, Providence, RI.

Monod, N. 2006a. An invitation to bounded cohomology. Pages 1183–1211 in *International Congress of Mathematicians. Vol. II.* European Mathematical Society, Zürich.

Monod, N. 2006b. Superrigidity for irreducible lattices and geometric splitting. *J. Amer. Math. Soc.*, **19**(4), 781–814.

Monod, N., and Shalom, Y. 2004. Cocycle superrigidity and bounded cohomology for negatively curved spaces. *J. Differential Geom.*, **67**(3), 395–455.

Monod, N., and Shalom, Y. 2006. Orbit equivalence rigidity and bounded cohomology. *Ann. of Math.*, 825–878.

Monod, N. 2001. *Continuous bounded cohomology of locally compact groups.* Lecture Notes in Mathematics, vol. 1758. Springer–Verlag, Berlin.

Monod, N. 2022. Lamplighters and the bounded cohomology of Thompson's group, *Geom. Funct. Anal.*, **32**(3), 662–675.

Moraschini, M. 2018. *On Gromov's theory of multicomplexes: the original approach to bounded cohomology and simplicial volume.* PhD thesis, Università di Pisa.

Mostow, G. D. 1968. Quasi-conformal mappings in n-space and the rigidity of hyperbolic space forms. *Inst. Hautes Études Sci. Publ. Math.*, 53–104.

Munkholm, H. J. 1980. Simplices of maximal volume in hyperbolic space, Gromov's norm, and Gromov's proof of Mostow's rigidity theorem (following Thurston). Pages 109–124 in *Topology Symposium, Siegen 1979 (Proc. Sympos., Univ. Siegen, Siegen, 1979).* Lecture Notes in Math., vol. 788. Springer, Berlin.

Myers, S. B., and Steenrod, N. E. 1939. The group of isometries of a Riemannian manifold. *Ann. of Math. (2)*, **40**(2), 400–416.

Noskov, G. A. 1991. Bounded cohomology of discrete groups with coefficients. *Leningrad Math. J.*, **2**(5), 1067–1084.

Okun, B., and Schreve, K. 2021. Torsion invariants of complexes of groups. *arXiv preprint arXiv:2108.08892.*

Ornstein, D. S., and Weiss, B. 1980. Ergodic theory of amenable group actions. I. The Rohlin lemma. *Bull. Amer. Math. Soc. (N.S.)*, **2**(1), 161–164.

Osin, D. 2016. Acylindrically hyperbolic groups. *Trans. Amer. Math. Soc.*, **368**(2), 851–888.

Ott, A. 2019. Transgression in bounded cohomology and a conjecture of Monod. *J. Topol. Anal.*, 1–40.

Pieters, H. 2018a. The boundary model for the continuous cohomology of $Isom^+(\mathbb{H}^n)$. *Groups Geom. Dyn.*, **12**(4), 1239–1263.

Pieters, H. 2018b. New bounds for the simplicial volume of complex hyperbolic surfaces. *arXiv preprint arXiv:1812.11541.*

Sauer, R. 2002. L^2-invariants of groups and discrete measured groupoids. PhD thesis, Westfälische Wilhelms–Universität Münster.

Sauer, R. 2009. Amenable covers, volume and L^2-Betti numbers of aspherical manifolds. *J. Reine Angew. Math.*, **636**, 47–92.

Savage, Jr., R. P. 1982. The space of positive definite matrices and Gromov's invariant. *Trans. Amer. Math. Soc.*, **274**(1), 239–263.

Schmidt, M. 2005. L^2-Betti numbers of $\mathcal{R}$-spaces and the integral foliated simplicial volume. PhD thesis, Westfälische Wilhelms–Universität Münster.

Sela, Zlil. 1992. Uniform embeddings of hyperbolic groups in Hilbert spaces. *Israel J. Math.*, **80**(1–2), 171–181.

Sisto, A. 2018. Contracting elements and random walks. *J. Reine Angew. Math.*, **2018**(742), 79–114.

Soma, T. 1997a. Bounded cohomology and topologically tame Kleinian groups. *Duke Math. J.*, **88**(2), 357–370.

Soma, T. 1997b. The zero-norm subspace of bounded cohomology. *Comment. Math. Helv.*, **72**(4), 582–592.

Strohm (=Löh), C. 2004. *The proportionality principle of simplicial volume*. Diploma thesis, Universität Münster. http://arxiv.org/abs/math/0504106.

Susse, T. 2015. Stable commutator length in amalgamated free products. *J. Topol. Anal.*, **7**(4), 693–717.

Thurston, W. P. 1979. *The geometry and topology of three-manifolds*. Princeton University Princeton, NJ.

Wolf, J. A. 2011. *Spaces of constant curvature*. 6th ed. AMS Chelsea Publishing, Providence, RI.

Yano, K. 1982. Gromov invariant and S^1-actions. *J. Fac. Sci. Univ. Tokyo Sect. IA Math.*, **29**(3), 493–501.

Zhuang, D. 2008. Irrational stable commutator length in finitely presented groups. *J. Mod. Dyn.*, **2**(3), 499–507.

Index

Index

Printed in the United States
by Baker & Taylor Publisher Services